Guide to Classical Physics

This is a "how to guide" for making introductory calculations in classical physics for undergraduates studying the subject.

The calculations are performed in Mathematica, and stress graphical visualization, units, and numerical answers. The techniques show the student how to learn the physics without being hung up on the math. There is a continuing movement to introduce more advanced computational methods into lower-level physics courses. Mathematica is a unique tool in that code is written as "human readable" much like one writes a traditional equation on the board.

The companion code for this book can be found here: https://physics.bu.edu/~rohlf/code.html

Key Features:

- Concise summary of the physics concepts
- Over 300 worked examples in Mathematica
- Tutorial to allow a beginner to produce fast results

James W. Rohlf is a Professor at Boston University. As a graduate student he worked on the first experiment to trigger on hadron jets with a calorimeter, Fermilab E260. His thesis (G. C. Fox, advisor, C. Barnes, R. P. Feynman, R. Gomez) used the model of Field and Feynman to compare observed jets from hadron collisions to that from electron-positron collisions and made detailed acceptance corrections to arrive at first the measurement of quark-quark scattering cross sections. His thesis is published in Nuclear Physics B171 (1980) 1. At the Cornell Electron Storage Rings, he worked on the discovery of the Upsilon (4S) resonance and using novel event shape variables developed by Stephen Wolfram and his thesis advisor, Geoffrey Fox. He performed particle identification of kaons and charmed mesons to establish the quark decay sequence, b –> c. At CERN, he worked on the discovery of the W and Z bosons and measurement of their properties. Presently, he is working on the Compact Muon Solenoid (CMS) experiment at the CERN Large Hadron Collider (LHC) which discovered the Higgs boson and is searching for new phenomena beyond the standard model.

Guide to Classical Physics

Using Mathematica for
Calculations and Visualizations

James W. Rohlf

CRC Press
Taylor & Francis Group
Boca Raton London New York

CRC Press is an imprint of the
Taylor & Francis Group, an **informa** business

Designed cover image: James W. Rohlf

First edition published 2025
by CRC Press
2385 NW Executive Center Drive, Suite 320, Boca Raton FL 33431

and by CRC Press
4 Park Square, Milton Park, Abingdon, Oxon, OX14 4RN

CRC Press is an imprint of Taylor & Francis Group, LLC

Library of Congress Cataloging-in-Publication Data
Names: Rohlf, James W., author.
Title: Guide to classical physics : using Mathematica for calculations and
visualizations / James W. Rohlf.
Description: First edition. | Boca Raton, FL : CRC Press, 2025.| Includes
bibliographical references and index. | Summary: "This is a "how to
guide" for making introductory calculations in classical physics for
undergraduates studying the subject. The calculations are performed in
Mathematica, and stress graphical visualization, units, and numerical
answers. The techniques show the student how to learn the physics
without being hung up on the math. There is a continuing movement to
introduce more advanced computational methods into lower-level physics
courses. Mathematica is a unique tool in that code is written as "human
readable" much like one writes a traditional equation on the board. The
companion code for this book can be found here:
https://physics.bu.edu/~rohlf/code.html"-- Provided by publisher.
Identifiers: LCCN 2024025160 | ISBN 9781032772417 (hbk) | ISBN
9781032769769 (pbk) | ISBN 9781003481980 (ebk)
Subjects: LCSH: Physics--Study and teaching (Secondary) | Mathematica
(Computer program language)
Classification: LCC QC30 .R637 2024 | DDC 530.071--dc23/eng/20240617
LC record available at https://lccn.loc.gov/2024025160

ISBN: 978-1-032-77241-7 (hbk)
ISBN: 978-1-032-76976-9 (pbk)
ISBN: 978-1-003-48198-0 (ebk)

DOI: 10.1201/9781003481980

Typeset in Nimbus Roman font
by KnowledgeWorks Global Ltd.

Publisher's note: This book has been prepared from camera-ready copy provided by the authors.

Contents

Preface

This is the first in a series of three books, which were written in reverse order, designed to help beginning students understand physics with the aid of Mathematica. The subjects covered are classical mechanics (this book), electricity and magnetism, and modern physics. Emphasis is placed on writing Wolfram language code that is as close as possible to standard notation found in textbooks. Units are meticulously accounted for in all calculations. Mathematica is used as a data base for physical constants, a calculator with units, an algebraic manipulator with math functions, an integrator, a differentiator, an equation solver, a series expander, a plotter, and more. It is important to note that Mathematica is not a substitute for learning the math, but rather a tool to greatly expand what you can do with your math knowledge. Often the computations illustrate the underlying physics in a way that is not otherwise possible.

The key to the uniqueness of Mathematica for this application is natural language input. Appendix A is intended to help a beginner get started immediately in Mathematica. One of the most important things to know at the outset is how physical constants are stored, retrieved and evaluated. A good general introduction, although somewhat abstract, is "An Elementary Introduction to the Wolfram Language," by Stephen Wolfram (Wolfram Media, 2023).

Mathematica has extensive in-line documentation of its over 6000 functions with a wide variety of example use cases. Hovering over a command will automatically link the support documentation material. Students should be encouraged to develop their own notebooks from which they can use as a reference to cut and paste as a quick starting point for new calculations.

The body of the text puts the calculations in physics context. The calculations are displayed as numbered examples, followed by executable code. The code for all the calculations and plots is downloadable.

CHAPTER 1

Units

Physical quantities have no meaning without their units. The scientific community uses the international system (SI) of units, which is based on the meter (m), second (s), and kilogram (kg).

1.1 LENGTH

The SI unit of length is the meter (m). The meter is defined as the distance that light travels in a vacuum in 1/299 792 458 of a second. It is obtained with the command Quantity["meter"] or by typing "meter" into the natural language box (see App. A).

Example 1.1 Get the meter unit.

```
In[2]:= Quantity["meter"]

Out[2]= 1 m
```

1.2 TIME

The SI unit of length is the second (s). The second is defined as an "atomic clock" by the transition frequency of the cesium atom. It is obtained with the command Quantity ["second"] or by typing "second" into the natural language box.

DOI: 10.1201/9781003481980-1

Example 1.2 Get the second unit.

```
In[3]:= Quantity["second"]

Out[3]= 1 s
```

1.3 MASS

The SI unit of mass is the kilogram (kg). The kilogram is defined via Planck's constant. It is obtained with the command Quantity["kilogram"] or by typing "kilogram" into the natural language box.

Example 1.3 Get the kilogram unit.

```
In[4]:= Quantity["kilogram"]

Out[4]= 1 kg
```

1.4 OTHER BASE UNITS

Many units needed in beginning physics are built from the base units of m, s, and kg; however, four more SI base units are needed to describe all of physics.

1.4.1 Temperature

Temperature on the absolute scale is measured in kelvin (K). Temperature is a measure of average motion under the condition of thermal equilibrium.

Example 1.4 Get the kelvin base unit.

```
In[6]:= Quantity["kelvin"]

Out[6]= 1 K
```

1.4.2 Electric Current

Electric current is measured in amperes (A). Current is a measure of electric charge times a velocity vector (Chap. 3). Electrical charge is the source of the electric force.

Example 1.5 Get the ampere base unit.

```
In[5]:= Quantity["ampere"]

Out[5]= 1 A
```

1.4.3 Quantity

The mole (mol) is a measure of the number of objects, as in "how many are there?" This usually refers to the number of atoms or molecules. One mole is a large (compared to 1) number of atoms because atoms are very small in volume (compared to 1 m^3).

Example 1.6 Get the mol base unit.

```
In[7]:= Quantity["mole"]

Out[7]= 1 mol
```

One mole of anything is exactly $6.02214076 \times 10^{23}$ of those things. This number is called Avogadro's number. Avogadro's number (N_0) is the number of objects in one mole,

$$N_0 = 6.02 \times 10^{23}.$$

1.4.4 Luminous Intensity

Luminous intensity is measured in candela (cd). Luminous intensity is a measure of the brightness of a source. The luminous intensity of a candle is about 1 cd. The luminous intensity of the sun is about 3×10^{27} cd.

Example 1.7 Get the candela base unit.

```
In[8]:= Quantity["candela"]

Out[8]= 1 cd
```

1.5   METRIC ABBREVIATIONS

Mathematica knows about metric abbreviations. The most common are milli (m) for 10^{-3}, micro (μ) for 10^{-6}, nano (n) for 10^{-9}, and pico (p) for 10^{-12}, plus kilo (k) for 10^3, mega (M) for 10^6, and giga (G) for 10^9.

The function UnitConvert (see App. A) attempts to convert a specified quantity to a specified unit. By default, the unit is given in SI base units.

Example 1.8 Convert 1 nm to m.

In[9]:= **UnitConvert[1. nm]**

Out[9]= $1. \times 10^{-9}$ m

Example 1.9 Convert 1 year to Ms.

In[10]:= **UnitConvert$\left[1.\ yr,\ M*s\right]$**

Out[10]= 31.536 M s

Example 1.10 Convert the mass of a basketball into grams.

In[11]:= **UnitConvert$\left[\ 1.\ \boxed{\text{NBA basketball} \text{ SPORT OBJECT}}\left[\boxed{mass}\right],\ g\right]$**

Out[11]= 623.69 g

1.6   DERIVED UNITS

Mathematica by default will report units in the base kg, m, and s. There are a number of frequently used derived units which are useful abbreviations and part of the language of physics. This includes the force unit, newton (N), the energy unit, joule (J), and the power unit, watt (W).

Example 1.11 Convert the newton unit into SI base units.

In[12]:= **UnitConvert$\left[N\right]$**

Out[12]= $1\ \text{kg}\,\text{m}/\text{s}^2$

Example 1.12 Convert the joule unit into SI base units.

In[13]:= **UnitConvert[J]**

Out[13]= $1 \text{ kg m}^2/\text{s}^2$

Example 1.13 Convert the watt unit into SI base units.

In[14]:= **UnitConvert[W]**

Out[14]= $1 \text{ kg m}^2/\text{s}^3$

1.7 NON-METRIC UNITS

Everyday life is full of non-metric units, requiring one to be able to convert between various units.

Example 1.14 Convert inches to m.

In[15]:= **UnitConvert[1. in]**

Out[15]= 0.0254 m

Astrophysicists love to use the parsec (pc) distance unit. One pc is the distance at which the astronomical unit (AU), or distance from the earth to the sun, subtends an angle of one arcsecond, or 1/3600 of one degree. One pc is the distance that light travels in 3.25 y.

Example 1.15 Convert 1 pc into m.

In[16]:= **UnitConvert[1. pc]**

Out[16]= $3.08568 \times 10^{16} \text{ m}$

Example 1.16 Convert 1 day into s.

In[17]:= **UnitConvert[1. days]**

Out[17]= $86400. \text{ s}$

Example 1.17 Convert 12 ounces (oz) into kg.

In[18]:= **UnitConvert[12. oz]**

Out[18]= 0.340194 kg

Example 1.18 Convert 60 miles per hour (mi/hr) into m/s.

In[19]:= **UnitConvert$\left[60. \dfrac{mi}{h}\right]$**

Out[19]= 26.8224 m/s

Example 1.19 Convert 1 teaspoon (tsp) into m^3.

In[20]:= **UnitConvert$\left[1. \text{ tsp}\right]$**

Out[20]= 4.92892 × 10^{-6} m^3

Mathematica will not compute a unit conversion if the units do not match. Suppose one tried to convert a teaspoon (tsp) into grams (g). This has no meaning because tsp is a volume unit and g is a mass unit.

Example 1.20 Try to convert 1 tsp into g.

In[21]:= **UnitConvert$\left[1. \text{ tsp, g}\right]$**

Out[21]= $Failed

A related calculation that would make sense is to convert 1 tsp of table salt (halite) into grams. Volume V times density ρ gives mass m,

$$V\rho = m.$$

Example 1.21 Calculate the mass of 1 tsp of table salt in g.

In[22]:= **UnitConvert$\left[(1. \text{ tsp}) \boxed{\text{halite } \text{MINERAL}} \left[\boxed{density}\right], \text{ g}\right]$**

Out[22]= 10.6958 g

1.8   SIGNIFICANT FIGURES

The number of digits displayed in a calculation is controlled by the function N, which attempts to give a result to a specified number of digits. For most beginning physics calculations, three figures is a good working point.

Example 1.22 Get the mass of the sun to three figures.

In[23]:= $N\left[\boxed{\textbf{Sun } \text{STAR}}\left[\boxed{mass}\right], 3\right]$

Out[23]= 1.99×10^{30} kg

1.9   PHYSICAL CONSTANTS

Mathematica knows about physical constants. Acceleration will be defined and used in Chap. 4.

Example 1.23 Get the acceleration of gravity.

In[24]:= $\textbf{UnitConvert}\left[1. \, g\right]$

Out[24]= 9.80665 m/s^2

The speed of light, together with the definition of the second, is used to define the meter (Sect. 1.1).

Example 1.24 Get the speed of light in vacuum.

In[25]:= $\textbf{UnitConvert[1. } c\textbf{]}$

Out[25]= 2.99792×10^8 m/s

Planck's constant, together with the definitions of the meter and second, is used to define the kg (Sect. 1.3).

Example 1.25 Get Planck's constant.

In[26]:= $\textbf{UnitConvert}\left[1. \, h\right]$

Out[26]= 6.62607×10^{-34} kg m^2/s

The universal gravitational constant will be discussed in Chap. 10.

Example 1.26 Get the universal gravitational constant.

In[27]:= **UnitConvert[G]**

Out[27]= 6.674×10^{-11} m^3/ (kg s^2)

The Boltzmann constant will be discussed in Chap. 12.

Example 1.27 Get the Boltzmann constant.

In[28]:= **UnitConvert[1. k]**

Out[28]= 1.38065×10^{-23} kg m^2/ (s^2 K)

Avogadro's number is the number of objects in 1 mole of those objects (Sect. 1.4.3).

Example 1.28 Get Avogadro's number.

In[29]:= **UnitConvert[$N_\ominus$]**

Out[29]= **602 214 076 000 000 000 000 000**

CHAPTER 2

Functions

Mathematica has over 6000 built in functions (see also App. A). Chapter 1 has already made use of the functions Quantity, UnitConvert. and N. All functions in Mathematica begin with a capital letter. Names of predefined functions cannot be used as user-defined variables.

2.1 TRIGONOMETRIC FUNCTIONS

The function Sin[θ] gives the sine of θ.

Example 2.1 Calculate $\sin\frac{\pi}{4}$.

In[1]:= **Sin**$\left[\frac{\pi}{4}\right]$

Out[1]= $\dfrac{1}{\sqrt{2}}$

The function Cos[θ] gives the cosine of θ.

Example 2.2 Calculate $\cos\frac{\pi}{3}$.

In[2]:= **Cos**$\left[\frac{\pi}{3}\right]$

Out[2]= $\dfrac{1}{2}$

Figure 2.1 plots the sine and cosine functions.
The identity,

$$\sin^2\theta + \cos^2\theta = 1,$$

DOI: 10.1201/9781003481980-2

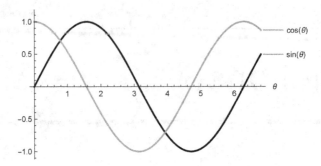

Figure 2.1 The functions $\sin\theta$ and $\cos\theta$ are plotted vs. θ.

is well-known. The function Simplify performs a sequence of algebraic and other transformations on and returns the simplest form it finds.

Example 2.3 Simplify $\sin^2\theta + \cos^2\theta$.

```
In[3]:= Simplify[Sin[θ]² + Cos[θ]²]
```

```
Out[3]= 1
```

Figure 2.2 shows a plot of $\sin^2\theta$ and $\cos^2\theta$.

The function Solve attempts to solve the system of equations or inequalities for the specified variables.

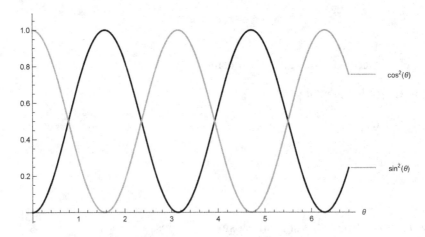

Figure 2.2 The functions $\sin^2\theta$ and $\cos^2\theta$ are plotted vs. θ.

Example 2.4 In Figure 2.1, find the places where the sine and cosine are equal.

In[4]:= **Solve[Sin[θ] == Cos[θ] && 0 < θ < 6, {θ}] // Simplify**

Out[4]= $\left\{\left\{\theta \to \dfrac{\pi}{4}\right\}, \left\{\theta \to \dfrac{5\pi}{4}\right\}\right\}$

The derived functions tangent, cotangent, secant, and cosecant given by Tan[θ], Cot[θ], Sec[θ], and Csc[θ], respectively. The definitions are

$$\tan\theta = \frac{\sin\theta}{\cos\theta},$$

$$\cot\theta = \frac{\cos\theta}{\sin\theta},$$

$$\sec\theta = \frac{1}{\cos\theta},$$

and

$$\csc\theta = \frac{1}{\sin\theta}.$$

2.2 DEGREES TO RADIANS

The preferred unit of angle is the dimensionless radian. One radian is $180/\pi$ degrees (°). The function that converts degrees into radians is Degree (no argument). Its equivalent symbol is the usual degree symbol.

2.3 INVERSE TRIGONOMETRIC FUNCTIONS

The inverse trigonometric functions are given by ArcSin[x], ArcCos[x], ArcTan[x], ArcCot[x], ArcSec[x], and ArcCsc[x].

Example 2.5 Get the degree symbol.

In[5]:= **Degree**

Out[5]= °

Example 2.6 Convert 30° to radians.

In[6]:= **30.°**

Out[6]= 0.523599

2.4 EXPONENTIAL FUNCTION

The exponential function is given by E (no argument).

Example 2.7 Get the exponential symbol.

In[7]:= **E**

Out[7]= e

The inverse of the exponential function is the natural logarithm (logarithm to base e). It is given by the function Log[x].

Example 2.8 Calculate ln e.

In[8]:= **Log[e]**

Out[8]= 1

2.5 USER-DEFINED FUNCTION

A user-defined function has the format $f[x_-]$.

Example 2.9 Particle physicists use an angular function called pseudorapidity (η), which is defined as $\eta = -\ln\tan\frac{\theta}{2}$. Define the function η and use it to calculate the pseudorapidity at $\theta = 5°$.

In[9]:= **$\eta[\theta_-] = -Log\left[Tan\left[\dfrac{\theta}{2}\right]\right]; \eta[5.°]$**

Out[9]= 3.1313

The pseudorapidity function is shown in Figure 2.3.

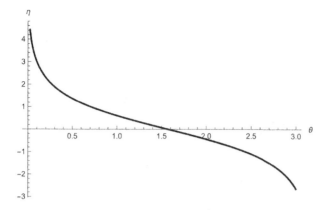

Figure 2.3 The function $\eta = \ln\tan\frac{\theta}{2}$ is plotted vs. θ.

2.6 DIFFERENTIATION

The function D gives the derivative of a function f with respect to the specified variable. The StandardForm notation is $\partial_x f$.

Example 2.10 Calculate the derivative of $\sin\theta$.

In[10]:= ∂_θ Sin[θ]

Out[10]= Cos[θ]

Example 2.11 Calculate the derivative of $\cos\theta$.

In[11]:= ∂_θ Cos[θ]

Out[11]= $-$Sin[θ]

Example 2.12 Calculate the derivative of $\tan x$.

In[12]:= ∂_θ Tan[θ]

Out[12]= Sec[θ]^2

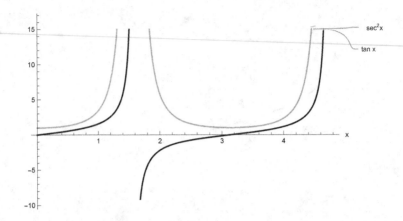

Figure 2.4   The function $\tan x$ and its derivatives are plotted vs. x.

The derivative is the slope of the function (Figure 2.4).

Example 2.13 Calculate the derivative of $\ln x$.

In[13]:= ∂_x Log[x]

Out[13]= $\dfrac{1}{x}$

The function $\ln x$ and its derivatives are shown in Figure 2.5.

Example 2.14 Calculate the derivative of η as defined in Ex. 2.9.

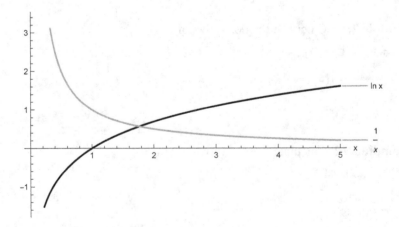

Figure 2.5   The function $\ln x$ and its derivatives are plotted vs. x.

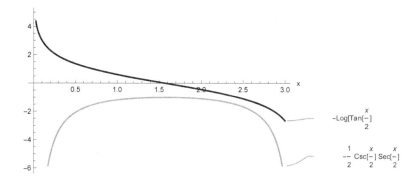

Figure 2.6 The function $\eta(x)$ and its derivatives are plotted vs. x.

In[14]:= $\partial_\theta \eta[\theta]$

Out[14]= $-\dfrac{1}{2} Csc\left[\dfrac{\theta}{2}\right] Sec\left[\dfrac{\theta}{2}\right]$

The function η and its derivatives are shown in Figure 2.6.

2.7 INTEGRATION

Mathematica will calculate both definite and indefinite integrals.

2.7.1 Indefinite Integrals

The function Integrate gives the indefinite integral over the specified variable.

Example 2.15 Display $\int dx f(x)$ in StandardForm.

In[15]:= **Integrate[f[x], x]**

Out[15]= $\int f[x]\, dx$

Example 2.16 Calculate $\int dx \sin x$.

In[16]:= $\int Sin[x]\, dx$

Out[16]= $-Cos[x]$

Example 2.17 Calculate $\int dx \ln x$.

In[17]:= $\int$ **Log[x] dx**

Out[17]= $-x + x$ **Log[x]**

2.7.2 Definite Integrals

The function Integrate[$f, \{x, x_{min}, x_{max}\}$ gives the indefinite integral $\int_{x_{min}}^{x_{max}} dx f(x)$.

Example 2.18 Calculate $\int_0^2 dx\, e^{-x} \cos x$.

In[18]:= $\int_{0}^{3}$ **e^{-x} Cos[x] dx**

N[%]

Out[18]= $\dfrac{e^3 - \text{Cos}[3] + \text{Sin}[3]}{2\, e^3}$

Out[19]= **0.528157**

The % is a shortcut to the last item that was executed.

A plot of $e^{-x} \cos x$ is shown in Figure 2.7. The definite integral is the area under the curve. When the curve goes negative, that portion of the area is negative.

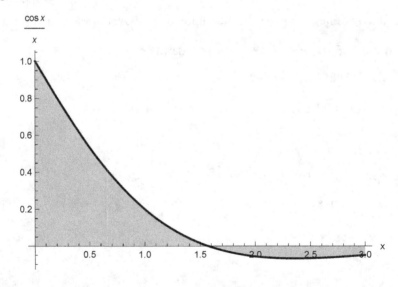

Figure 2.7 The function $e^{-x} \cos x$ is plotted vs. x. The definite integral is the area under the curve.

CHAPTER 3

Vectors

Vectors in Mathematica are represented with curly brackets { }.

3.1   UNIT VECTORS

The unit vectors in Cartesian coordinates are represented by {1,0,0}, {0,1,0}, and {0,0,1}.

Example 3.1 Define the unit vectors $\hat{\mathbf{x}}$, $\hat{\mathbf{y}}$, and $\hat{\mathbf{z}}$.

```
In[1]:= x̂ = {1, 0, 0}
        ŷ = {0, 1, 0}
        ẑ = {0, 0, 1}

Out[1]= {1, 0, 0}

Out[2]= {0, 1, 0}

Out[3]= {0, 0, 1}
```

A vector is denoted by its (x, y, z) components. For example, a vector **a** having components (a_x, a_y, a_z) is expressed in terms of unit vectors as

$$\mathbf{a} = a_x\hat{\mathbf{x}} + a_y\hat{\mathbf{y}} + a_z\hat{\mathbf{z}}.$$

In Mathematica, the expression a[[n]] selects the $x-$, $y-$, or $z-$component for $n = 1$, 2, or 3, respectively.

Example 3.2 Select the $x-$, $y-$, and $z-$components of the vector **a**.

DOI: 10.1201/9781003481980-3

In[4]:= `<< Notation`

`Symbolize[ParsedBoxWrapper[SubscriptBox["_", "_"]]]`

$a = \{a_x, a_y, a_z\}; a[[1]]$

$a[[2]]$

$a[[3]]$

Out[6]= a_x

Out[7]= a_y

Out[8]= a_z

Technically, the use of a subscript can produce an illegal variable in Mathematica. For example, a subscripted variable cannot be cleared, so one needs to use a little bit of caution in its use. They have a huge advantage, however, in making the code readable. Subscripted variables can cause confusion with a number of vector operations. The first two lines of code in Ex. 3.2 protect the use of subscripts in this example the and ones to follow. They produce no output.

3.2 SCALAR-PRODUCT

The scalar-product of two vectors is commonly known as the dot product. The dot product of vectors **a** and **b** (**a** · **b**) is defined as

$$\mathbf{a} \cdot \mathbf{b} = a_x b_x + a_y b_y + a_z b_z.$$

The function for the dot product is Dot[a,b] for vectors a and b and its shortcut is a period between the two vectors.

Example 3.3 Calculate **a** · **b**.

In[9]:= $a = \{a_x, a_y, a_z\}; b = \{b_x, b_y, b_z\};$

`a.b`

Out[10]= $a_x b_x + a_y b_y + a_z b_z$

The order of the dot-product does not matter,

$$\mathbf{a} \cdot \mathbf{b} = \mathbf{b} \cdot \mathbf{a}$$

Example 3.4 Compare $\mathbf{a} \cdot \mathbf{b}$ with $\mathbf{b} \cdot \mathbf{a}$,

In[11]:= **a.b == b.a**

Out[11]= True

The double equal sign in Ex. 3.4 does a logical comparison.

Unit vectors are orthogonal, meaning that the dot-product of any two is one if they are identical and zero if they are different.

Example 3.5 Calculate $\hat{\mathbf{x}} \cdot \hat{\mathbf{x}}$, $\hat{\mathbf{y}} \cdot \hat{\mathbf{y}}$, and $\hat{\mathbf{z}} \cdot \hat{\mathbf{z}}$.

In[12]:= **x̂.x̂**
      ŷ.ŷ
      ẑ.ẑ

Out[12]= 1

Out[13]= 1

Out[14]= 1

Example 3.6 Calculate $\hat{\mathbf{x}} \cdot \hat{\mathbf{y}}$, $\hat{\mathbf{x}} \cdot \hat{\mathbf{z}}$, and $\hat{\mathbf{y}} \cdot \hat{\mathbf{z}}$.

In[15]:= **x̂.ŷ**
      x̂.ẑ
      ŷ.ẑ

Out[15]= 0

Out[16]= 0

Out[17]= 0

The dot-product of a vector with a unit vector selects the corresponding component of the vector.

Example 3.7 Calculate $\mathbf{a} \cdot \hat{\mathbf{x}}$, $\mathbf{a} \cdot \hat{\mathbf{y}}$, and $\mathbf{a} \cdot \hat{\mathbf{z}}$,

In[18]:= **a.x̂**

a.ŷ

a.ẑ

Out[18]= a_x

Out[19]= a_y

Out[20]= a_z

3.3 SQUARE OF THE MAGNITUDE

The square of the magnitude of a vector is the dot-product of the vector with itself,

$$a^2 = \mathbf{a} \cdot \mathbf{a}.$$

3.4 COSINE RULE FOR DOT PRODUCT

The cosine rule for the dot product states that

$$\mathbf{a} \cdot \mathbf{b} = ab\cos\theta,$$

where θ is the angle between $\mathbf{a}$ and $\mathbf{b}$. Any two vectors will lie in a plane so one may choose coordinates such that $\mathbf{a}$ and $\mathbf{b}$ are in the $x - y$ plane and $\mathbf{b}$ is in the x−direction (Figure 3.1). Then it is easily seen that

$$\mathbf{a} \cdot \mathbf{b} = a_x b = \frac{a_x}{a}ab = ab\cos\theta.$$

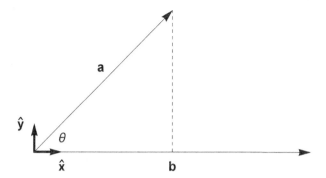

Figure 3.1   The coordinate system is chosen such that **a** and **b** are in the $x-y$ plane and **b** is in the x-direction. The vector **a** points in an arbitrary direction given by the angle θ.

Example 3.8 For vectors **a** and **b** in the $x-y$ plane and $\mathbf{b} = b\hat{\mathbf{x}}$, calculate $\frac{\mathbf{a}\cdot\mathbf{b}}{ab}$.

In[21]:= **\$Assumptions = b$_x$ > 0;**

$$\frac{\mathbf{a.b}}{\sqrt{\mathbf{a.a}}\ \sqrt{\mathbf{b.b}}} \ \text{/. } \{b_y \to 0, \ b_z \to 0, \ a_z \to 0\} \text{ // Simplify}$$

Out[22]= $\dfrac{a_x}{\sqrt{a_x^2 + a_y^2}}$

This leads to the law of cosines which relates the length of one side of a triangle to the lengths of the other two sides and the angle between them (Figure 3.2),

$$c^2 = (\mathbf{a-b})\cdot(\mathbf{a-b}) = a^2 + b^2 - 2\mathbf{a}\cdot\mathbf{b},$$

or

$$c = \sqrt{a^2 + b^2 - 2ab\cos\theta}.$$

This is an important result which is worth understanding thoroughly.

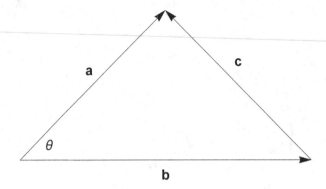

Figure 3.2 The coordinate system is chosen such that the vector **b** is in the x–direction. The vector **a** points in an arbitrary direction.

Example 3.9 Generate two random vectors and calculate the angle between them.

In[23]:= `r = RandomReal[1, {2, 3}]; A = r[[1]]`
`B = r[[2]]`
$s = Solve\left[A.B == \sqrt{A.A} \; \sqrt{B.B} \; Cos[\theta], \; \theta\right]$

Out[23]= `{0.725812, 0.344226, 0.183683}`

Out[24]= `{0.0460314, 0.503639, 0.800472}`

Out[25]= $\{\{\theta \rightarrow -1.10015\}, \; \{\theta \rightarrow 1.10015\}\}$

The function RandomReal[x_{max}, {n_1, n_2}] gives a random number in the range 0 to x_{max} in an array $n_1 \times n_2$.

The default unit of an angle is a dimensionless quantity that is referred to as the radian. It is kind of weird to give a dimensionless quantity a name that sounds like a unit, but that is what has been done. If you see the word radian you can just cross it out. It is there to remind you that the angle is NOT given in degrees. The function Degree gives the number of radians per degree.

Example 3.10 Convert the positive angle in Ex. 3.9 into degrees.

In[26]:= $\dfrac{\theta \; /. \; s[[2]]}{Degree}$

Out[26]= `63.034`

Note that the solution of Ex. 3.9 is reported as a List. In Ex. 3.10, θ is evaluated by using the second solution ($s[[2]]$).

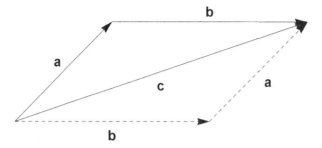

Figure 3.3 Two vectors **a** and **b** are added by placing the "tail" of **b** at the same place as the "arrow" of **a**.

3.5 ADDING AND SUBTRACTING VECTORS

The addition of two vectors is visualized by placing the "tail" of one vector and the "arrow" of the other vector at the same point. The addition may be done in any order. This is illustrated in Figure 3.3.

The subtraction of two vectors is visualized by placing the "tails" at the same point (Figure 3.2).

3.6 VECTOR PRODUCT

The vector product of two vectors is also known as the cross product. The cross product of vectors **A** and **B** is written as $\mathbf{A} \times \mathbf{B}$.

The function for the cross product is Cross[a,b] for vectors a and b and its shortcut is a multiplication sign ($\times$) between the two vectors.

The cross product of a vector with itself is zero.

Example 3.11 Calculate $\hat{\mathbf{x}} \times \hat{\mathbf{x}}$, $\hat{\mathbf{y}} \times \hat{\mathbf{y}}$, and $\hat{\mathbf{z}} \times \hat{\mathbf{z}}$.

```
In[27]:= x̂ × x̂
         ŷ × ŷ
         ẑ × ẑ

Out[27]= {0, 0, 0}

Out[28]= {0, 0, 0}

Out[29]= {0, 0, 0}
```

Unit vectors are orthogonal, meaning that the cross product of any two produces the third. The unit vectors are right-handed, meaning $\hat{\mathbf{x}} \times \hat{\mathbf{y}}$ makes $\hat{\mathbf{z}}$ (not $-\hat{\mathbf{z}}$).

Example 3.12 Calculate $\hat{\mathbf{x}} \times \hat{\mathbf{y}}$, $\hat{\mathbf{y}} \times \hat{\mathbf{z}}$, and $\hat{\mathbf{z}} \times \hat{\mathbf{x}}$.

In[30]:= $\hat{\mathbf{x}} \times \hat{\mathbf{y}}$

$\hat{\mathbf{y}} \times \hat{\mathbf{z}}$

$\hat{\mathbf{z}} \times \hat{\mathbf{x}}$

Out[30]= {0, 0, 1}

Out[31]= {1, 0, 0}

Out[32]= {0, 1, 0}

Example 3.13 Calculate $\hat{\mathbf{y}} \times \hat{\mathbf{x}}$, $\hat{\mathbf{z}} \times \hat{\mathbf{y}}$, and $\hat{\mathbf{x}} \times \hat{\mathbf{z}}$.

In[33]:= $\hat{\mathbf{y}} \times \hat{\mathbf{x}}$

$\hat{\mathbf{z}} \times \hat{\mathbf{y}}$

$\hat{\mathbf{x}} \times \hat{\mathbf{z}}$

Out[33]= {0, 0, -1}

Out[34]= {-1, 0, 0}

Out[35]= {0, -1, 0}

The unit vector cross-products give

$$\mathbf{a} \times \mathbf{b} = (a_y b_z - a_z b_y)\hat{\mathbf{x}} + (a_z b_x - a_x b_z)\hat{\mathbf{y}} + (a_x b_y - a_y b_x)\hat{\mathbf{x}}.$$

This is often written as a determinant,

$$\mathbf{a} \times \mathbf{b} = \begin{vmatrix} \hat{\mathbf{x}} & \hat{\mathbf{y}} & \hat{\mathbf{z}} \\ a_x & a_y & a_z \\ b_x & b_y & b_z \end{vmatrix} = \hat{\mathbf{x}} \begin{vmatrix} a_y & a_z \\ b_y & b_z \end{vmatrix} - \hat{\mathbf{y}} \begin{vmatrix} a_x & a_z \\ b_x & b_z \end{vmatrix} + \hat{\mathbf{z}} \begin{vmatrix} a_x & a_y \\ b_x & b_y \end{vmatrix},$$

where

$$\begin{vmatrix} a_y & a_z \\ b_y & b_z \end{vmatrix} = a_y b_z - a_z b_y,$$

etc.

Example 3.14 Calculate $\mathbf{a} \times \mathbf{b}$.

In[36]:= **a** × **b**

Out[36]= $\left\{ -a_z b_y + a_y b_z, \ a_z b_x - a_x b_z, \ -a_y b_x + a_x b_y \right\}$

3.7 SINE RULE FOR CROSS PRODUCT

Using the same coordinates as Figure 3.1, the cross-product is

$$\mathbf{a} \times \mathbf{b} = (a_x\hat{\mathbf{x}} + a_y\hat{\mathbf{y}}) \times (b_x\hat{\mathbf{x}}) = -a_y b_x \hat{\mathbf{z}} = (-a)b_x \frac{a_y}{a}\hat{\mathbf{z}} = -ab_x \sin\theta\,\hat{\mathbf{z}},$$

where a is the magnitude of the vector $\mathbf{a}$. The magnitude of the cross-product is

$$|\mathbf{a} \times \mathbf{b}| = ab\sin\theta.$$

Example 3.15 For vectors $\mathbf{a}$ and $\mathbf{b}$ in the $x - y$ plane and $\mathbf{b} = b\hat{\mathbf{x}}$, calculate $\frac{|\mathbf{a}\times\mathbf{b}|}{ab}$.

In[37]:= `$Assumptions = {bₓ > 0, a_y > 0};`

$$\frac{\sqrt{(\mathbf{a} \times \mathbf{b}) \cdot (\mathbf{a} \times \mathbf{b})}}{\sqrt{\mathbf{a}.\mathbf{a}}\ \sqrt{\mathbf{b}.\mathbf{b}}}\ \texttt{/.}\ \{b_y \to 0,\ b_z \to 0,\ a_z \to 0\}\ \texttt{// Simplify}$$

Out[38]= $\dfrac{a_y}{\sqrt{a_x^2 + a_y^2}}$

3.8 VECTOR EQUATIONS

A vector equation is an extremely convenient short-hand notation for writing three equations at once. For example, writing

$$\mathbf{c} = \mathbf{a} + \mathbf{b}$$

means

$$c_x = a_x + b_x,$$

$$c_y = a_y + b_y,$$

and

$$c_z = a_z + b_z.$$

3.9 TRIPLE VECTOR PRODUCT

The triple vector product, $\mathbf{a} \times (\mathbf{b} \times \mathbf{c})$, gives the "bac-cab" identity,

$$\mathbf{a} \times (\mathbf{b} \times \mathbf{c}) = \mathbf{b}(\mathbf{a} \cdot \mathbf{c}) - \mathbf{c}(\mathbf{a} \cdot \mathbf{b}).$$

Example 3.16 Calculate $\mathbf{a} \times (\mathbf{b} \times \mathbf{c})$ and verify the bac-cab identity.

```
In[39]:= c = {cx, cy, cz};
         a × (b × c)
         a × (b × c) == b (a.c) - c (a.b) // Simplify
```

$$Out[40]= \{ -a_y b_y c_x - a_z b_z c_x + a_y b_x c_y + a_z b_x c_z,$$
$$a_x b_y c_x - a_x b_x c_y - a_z b_z c_y + a_z b_y c_z,$$
$$a_x b_z c_x + a_y b_z c_y - a_x b_x c_z - a_y b_y c_z \}$$

```
Out[41]= True
```

CHAPTER 4

Motion in One Dimension

The tools for describing motion are the position, velocity, and acceleration vectors. Equations of motion are defined and derived from a position vector, which depends on time.

4.1 POSITION

4.1.1 Role of Time

Time is an interesting and sophisticated concept in physics. Time is related to motion. The day, for example, is defined by the rotation of the earth about its axis. In special relativity, space and time are deeply connected through the speed of light. Kip Thorne (2017 Nobel laureate) once said "Time is defined so that motion appears simple." In simple terms, time is used as a parameter (the independent variable) on which the position (the dependent variable) depends.

The motion of an object is described by specifying the position vector $\mathbf{x}(t)$ as a function of time,

$$\mathbf{x}(t) = x(t)\,\hat{\mathbf{x}}.$$

A number of related quantities may be derived from $x(t)$. As an example, suppose that

$$x = 30 \text{ m} - \left(40\frac{\text{m}}{\text{s}}\right)t + \left(2\frac{\text{m}}{\text{s}^3}\right)t^3.$$

The motion is along a straight line (in either direction). There is only one unit vector, which is common to all vector quantities. One must keep in mind however, that x or any of the derived quantities could be positive or negative.

DOI: 10.1201/9781003481980-4

Notice that in defining the function, the units of each term must match. One must take care to code the function in Mathematica with the correct units, so that the units of all derived quantities will automatically be calculated. This provides a powerful check of the correctness of the calculations.

Example 4.1 Define the position vector.

$$\text{In[1]:= } x[t\_] = 30 \text{ m} - 40 \; \frac{m}{s} \, t + 2 \; \frac{m}{s^3} \, t^3$$

$$\text{Out[1]= } t \left(-40 \text{ m/s} \right) + t^3 \left(2 \text{ m/s}^3 \right) + 30 \text{ m}$$

4.1.2 Displacement

The displacement vector, $\Delta\mathbf{x} = \Delta x \, \hat{\mathbf{x}}$, denotes a change in position. It is calculated by subtracting the initial position x_1 from the final position x_2,

$$\Delta x = x_2 - x_1.$$

As noted above, Δx could be positive or negative. For projectile motion, the displacement vector lies in a plane, while in three dimensions, the displacement vector can point in any direction. This is covered in Chap. 5.

Example 4.2 Use the position vector to calculate the displacement for the time interval $t_1 = 0$ s to $t_2 = 4$ s.

$$\text{In[3]:= } \frac{x[t_2] - x[t_1]}{t_2 - t_1}$$

$$\text{Out[3]= } -8 \text{ m/s}$$

Notice that in going from one Mathematica input cell (Ex. 4.1) to the next (Ex. 4.2), the function stays defined unless it is cleared.

4.2 VELOCITY

The velocity vector is the change in position vector per time.

4.2.1 Average Velocity

The average velocity vector $\bar{\mathbf{v}} = \bar{v} \, \hat{\mathbf{x}}$, over a time interval

$$\Delta t = t_2 - t_1,$$

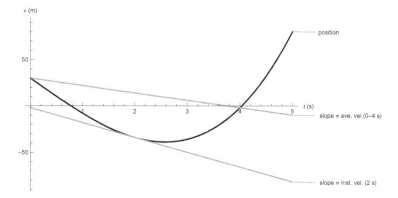

Figure 4.1 The position is given by $x = 30 \text{ m} - \left(40\frac{\text{m}}{\text{s}}\right)t + \left(2\frac{\text{m}}{\text{s}^3}\right)t^3$. The instantaneous velocity at any time is the slope of the tangent to the position curve. The average velocity over any time interval $t_2 - t_1$ is the slope of the line that connects the two times t_1 and t_2.

is the displacement vector that occurred during that time interval divided by Δt,

$$\bar{v} = \frac{\Delta x}{\Delta t} = \frac{x_2 - x_1}{t_2 - t_1}.$$

The average velocity is the slope of the line that connects the two times that define the position interval (Figure 4.5).

Example 4.3 Calculate the average velocity for the time interval $t_i = 0$ s to $t_f = 4$ s.

In[3]:= $\dfrac{\text{x}[\text{t}_2] - \text{x}[\text{t}_1]}{\text{t}_2 - \text{t}_1}$

Out[3]= -8 m/s

4.2.2 Instantaneous Velocity

The instantaneous velocity vector, $\mathbf{v} = v\,\hat{\mathbf{x}}$, is given by average velocity in the limit that the time interval of the displacement becomes infinitesimally small,

$$v = \lim_{\Delta t \to 0} \frac{\Delta x}{\Delta t} = \frac{dx}{dt}.$$

This also means that

$$x = \int dt\, v(t),$$

and

$$x_2 = x_1 + \int_{t_1}^{t_2} dt\, v(t).$$

The instantaneous velocity for the $x(t)$ of Sect. 4.1.1 is

$$v = \frac{dx}{dt} = \frac{d}{dt}\left[30\text{ m} - \left(40\frac{\text{m}}{\text{s}}\right)t + \left(2\frac{\text{m}}{\text{s}^3}\right)t^3\right] = -40\frac{\text{m}}{\text{s}} + \left(6\frac{\text{m}}{\text{s}^3}\right)t^2.$$

Example 4.4 Calculate the instantaneous velocity at $t = 2$ s.

In[4]:= **v[t_] = ∂ₜ x[t]; v[2 s]**

Out[4]= −16 m/s

The instantaneous velocity is the slope of the line that is tangent to the position curve (Figure 4.5). For a tiny enough time interval, the average velocity will be equal to the instantaneous velocity.

Example 4.5 Calculate the average velocity for $t_1 = (2 - 0.001)$ s and $t_2 = (2 + 0.001)$ s.

In[5]:= **t₁ = (2 - 0.001) s; t₂ = (2 + 0.001) s;** $\dfrac{\text{x[t₂] - x[t₁]}}{\text{t₂ - t₁}}$

Out[5]= −16. m/s

4.2.3 Speed

The language of physics is careful to distinguish between the velocity **v**, which is a vector, and its magnitude of **v** which is called the speed (s). The instantaneous speed is given by

$$s = \sqrt{\mathbf{v} \cdot \mathbf{v}} = |v|.$$

Note that this is not the same v, even in one dimension, because v could be negative, whereas the speed is always positive.

The average speed $\bar{s}$ over any time interval $\Delta t = t_2 - t_1$ is the total distance traveled (d) divided by the Δt. The total distance traveled depends only on the speed and not on the direction,

$$d = \int_{t_1}^{t_2} dt\, |v|.$$

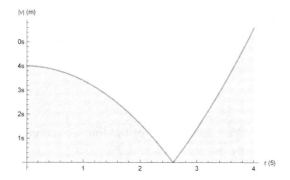

Figure 4.2   The distance traveled is the area under the plot of $|v|$.

In a plot of $|v|$ vs. t, d is the area under the curve (Figure 4.2). This gives

$$\bar{s} = \frac{d}{\Delta t} = \frac{\int_{t_1}^{t_2} dt \, |v|}{t_2 - t_1}.$$

To calculate the total distance traveled, one needs to integrate the magnitude of the instantaneous velocity (the instantaneous speed) over the time integral $t_f - t_i$.

Example 4.6   Calculate the average speed for $t_i = 0$ s to $t_f = 4$ s.

In[6]:= $\mathbf{t_1 = 0\ s;\ t_2 = 4\ s;}$ $\dfrac{\int_{t_1}^{t_2} \sqrt{v[t]^2}\ dt}{t_2 - t_1}$

$\mathbf{N[\%]}$

Out[6]= $\dfrac{8}{9}\left(-9 + 10\ \sqrt{15}\right)$ m/s

Out[7]= 26.4265 m/s

4.3   ACCELERATION

The acceleration vector is the change in velocity vector per time.

4.3.1 Average Acceleration

The average acceleration vector, $\bar{\mathbf{a}} = \bar{a}\,\hat{\mathbf{x}}$, over a time interval Δt is the change in velocity vector that occurred during that time interval divided by Δt,

$$\bar{a} = \frac{\Delta v}{\Delta t} = \frac{v_f - v_i}{t_f - t_i}.$$

The average acceleration is the slope of the line that connects the two times that define the velocity interval (Figure 4.3).

Example 4.7 Calculate the average acceleration for $t_i = 0$ s to $t_f = 2$ s.

In[8]:= $\dfrac{\mathbf{v[t_2] - v[t_1]}}{\mathbf{t_2 - t_1}}$

Out[8]= 24 m/s^2

4.3.2 Instantaneous Acceleration

The instantaneous acceleration vector, $\mathbf{a} = a\,\hat{\mathbf{x}}$, is given by average acceleration in the limit that the time interval of the change in velocity becomes infinitesimally small,

$$a = \lim_{\Delta t \to 0} \frac{\Delta v}{\Delta t} = \frac{dv}{dt}.$$

This also means that

$$v = \int dt\, a(t),$$

and

$$v_2 = v_1 + \int_{t_1}^{t_2} dt\, a(t).$$

The instantaneous acceleration for the $v(t)$ of Sect. 4.2.2 is

$$a = \frac{dv}{dt} = \frac{d}{dt}\left[-40\frac{m}{s} + \left(6\frac{m}{s^3}\right)t^2\right] = \left(12\frac{m}{s^3}\right)t.$$

Example 4.8 Calculate the instantaneous acceleration at $t = 2$ s.

In[9]:= $\mathbf{a[t\_] = \partial_t\, v[t]; a[2\ s]}$

Out[9]= 24 m/s^2

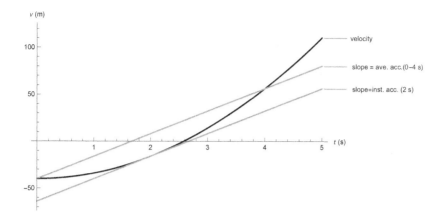

Figure 4.3 The velocity is given by $v = -40\frac{m}{s} + \left(6\frac{m}{s^3}\right)t^2$. The instantaneous acceleration at any time is the slope of the tangent to the velocity curve. The average acceleration over any time interval $t_2 - t_1$ is the slope of the line that connects the two times t_2 and t_1.

The instantaneous velocity is the slope of the line that is tangent to the velocity curve (Figure 4.3). For a tiny enough time interval, the average acceleration will be equal to the instantaneous acceleration.

Example 4.9 Calculate the average acceleration in the interval $t_i = (2 - 0.001)$ s to $t_f = (2 + 0.001)$ s.

```
In[10]:= t₁ = (2 - 0.001) s; t₂ = (2 + 0.001) s;   v[t₂] - v[t₁]
                                                   ───────────
                                                     t₂ - t₁

Out[10]= 24. m/s²
```

4.4 ANOTHER EXAMPLE

The code from the above examples can be used as a template to calculate $d, \bar{v}, v, \bar{a}$ and a from any position function $x(t)$. Suppose that

$$x(t) = (4 \text{ m})\cos\frac{t}{s} - \left(0.3\,\frac{m}{s^2}\right)t^2.$$

Figure 4.4 shows a plot of $x(t)$.

Example 4.10 Calculate $\Delta x, \bar{v}$, and $\bar{a}$ for $t_1 = 0$ s to $t_2 = 10$ s, and v and a at time $t_f = 2$ s.

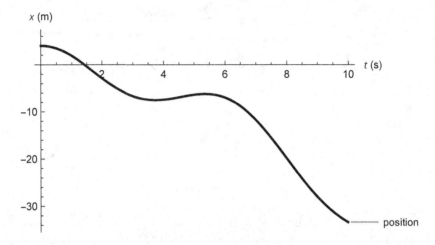

Figure 4.4 The position is given by $x(t) = (4\text{ m})\cos\frac{t}{s} - \left(0.3\,\frac{m}{s^2}\right)t^2$.

$\text{In[11]:= } \mathbf{x[t\_] = (4\ m)\ Cos\left[\dfrac{t}{s}\right] - \left(.3\ \dfrac{m}{s^2}\right)t^2;}$

$\mathbf{t_1 = 0\ s;}$
$\mathbf{t_2 = 10\ s;}$
$\mathbf{x[t_2] - x[t_1]}$
$\mathbf{\dfrac{x[t_2] - x[t_1]}{t_2 - t_1}}$

$\mathbf{v[t\_] = \partial_t\,x[t];\ \dfrac{v[t_2] - v[t_1]}{t_2 - t_1}}$

$\mathbf{\dfrac{\int_{t_1}^{t_2} \sqrt{v[t]^2}\ dt}{t_2 - t_1}}$

$\mathbf{v[2\ s]}$
$\mathbf{a[t\_] = \partial_t\,v[t];\ a[2\ s]}$

$\text{Out[12]= } -37.3563\ m$

$\text{Out[13]= } -3.73563\ m/s$

$\text{Out[14]= } -0.382392\ m/s^2$

$\text{Out[15]= } 3.99475\ m/s$

$\text{Out[16]= } -4.83719\ m/s$

$\text{Out[17]= } 1.06459\ m/s^2$

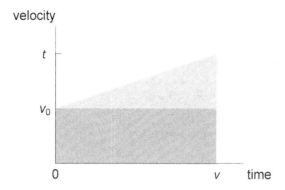

Figure 4.5 In a plot of velocity vs. time for constant acceleration, the area under the curve is the change in position given by the area of the rectangle ($v_0 t$) plus the area of the triangle whose height is the change in velocity (at) and whose base is the time (t).

4.5 CONSTANT ACCELERATION

The special case of constant acceleration occurs frequently and is worth a detailed analysis. Constant downward acceleration together with horizontal motion gives projectile motion which is discussed in Chap 5. For constant acceleration,

$$v = a \int dt = v_0 + at,$$

where the constant of integration v_0 is the velocity at $t = 0$. The position is

$$x = \int dt \, v = x_0 + v_0 t + \frac{1}{2} at^2,$$

where the constant of integration x_0 is the position at $t = 0$. The quantity $x - x_0$ is given by the area under the curve of a velocity vs. time plot where the area of the rectangle is $v_0 t$ and the area of the triangle is $\frac{1}{2}(t)(at)$.

Example 4.11 An object has an initial velocity of $v_0 = 4$ m/s and a constant acceleration of $a = 2.5$ m/s². How far does it travel in 2.2 s and what is its resulting velocity?

In[18]:= **ClearAll["Global`*"];**

<< Notation`

Symbolize[ParsedBoxWrapper[SubscriptBox["_", "_"]]]

$$v_\theta = 4 \ \frac{m}{s} ; \ a = 2.5 \ \frac{m}{s^2} ;$$

$$v_\theta \ t + \frac{1}{2} \ a \ t^2 \ /. \ t \to 2.2 \ s$$

$$v_\theta + a \ t \ /. \ t \to 2.2 \ s$$

Out[22]= 14.85 m

Out[23]= 9.5 m/s

4.5.1 Equation with No Time

One may eliminate t from the constant acceleration motion equations.

Example 4.12 Eliminate t from the motion equations.

In[24]:= **ClearAll["Global`*"];**

$$\textbf{Solve}\left[v == v_\theta + a \ t \ \&\& \ x == x_\theta + v_\theta \ t + \frac{1}{2} \ a \ t^2, \ \{a, \ t\}\right]$$

Out[25]= $\left\{\left\{a \to \dfrac{v^2 - v_\theta^2}{2 \ (x - x_\theta)}, \ t \to \dfrac{2 \ (x - x_\theta)}{v + v_\theta}\right\}\right\}$

It is seen that

$$v^2 = v_0^2 + 2a(x - x_0).$$

Example 4.13 Suppose the initial velocity of an automobile was 60 mi/hr. What acceleration is needed to bring the automobile to rest in 200 ft?

In[26]:= **ClearAll["Global`*"];**

$$v_\theta = 60 \ \frac{mi}{h} ; \ v = 0;$$

$$a = \textbf{UnitConvert}\left[\frac{v^2 - v_\theta^2}{2 \ (x - x_\theta)} \ /. \ x \to x_\theta + 200. \ ft\right]$$

Out[28]= -5.90093 m/s^2

4.5.2 Average Velocity

Consider an object with initial speed v_0 that starts at $x_0 = 0$ and accelerates with constant acceleration a for a time t. The speed is

$$v = v_0 + at,$$

and the displacement is

$$d = v_0 t + \frac{1}{2} a t^2.$$

and the speed is

$$v = v_0 + at.$$

The average velocity is

$$\bar{v} = \frac{d}{t} = v_0 + \frac{at}{2} = v_0 + \frac{v - v_0}{2} = \frac{v + v_0}{2}.$$

Example 4.14 The space shuttle accelerates from zero to 8000 m/s in 8.5 min. Assuming constant acceleration, how far did it travel?

In[29]:= $\mathbf{t = 8.5 \, min; \, v_\theta = 0; \, v = 8000 \, \frac{m}{s};} \quad \frac{\mathbf{v + v_\theta}}{\mathbf{2}} \mathbf{t}$

Out[29]= 2.04×10^6 m

4.5.3 Free Fall

The acceleration of gravity is denoted with the symbol g (see Ex. 1.23). The constant g is defined as a positive number, so that for free-fall motion,

$$a = -g.$$

This gives

$$v = v_0 - gt,$$

and

$$x = x_0 + v_0 t - \frac{1}{2} g t^2.$$

Example 4.15 An object is launched upward at an unknown speed. It comes to rest at an unknown height 3 s later. Calculate the initial speed and the height reached from its starting position.

In[20]:= `ClearAll["Global`*"];`

`Solve[0 == v` $_0$ ` - g t && x == v` $_0$ ` t - 0.5 g t` ^{2} `, {x, v` $_0$ `}] /.`

` t → 3 s`

Out[31]= $\{\{x → 44.1299\, m, v_0 → 29.42\, m/s\}\}$

4.6 COMPARING THE MOTION OF TWO OBJECTS

Consider two objects that are described by different equations of motion.

Example 4.16 Two trains have the same position at $t = 0$. Train 1 moves at a constant speed of 40 m/s and train 2 accelerates from rest at 1 m/s^2. How long does it take for train 2 to catch up to train 1 and at what distance does this happen?

In[32]:= `s = Solve[v t ==` $\frac{1}{2}$ `a t` ^{2} `, {t}] /. v → 40` $\frac{m}{s}$

`(t /. s⟦2⟧) v /.` $\left\{v → 40\, \frac{m}{s}, a → 1\, \frac{m}{s^2}\right\}$

Out[32]= $\left\{\{t → 0\}, \left\{t → \frac{80\, m/s}{a}\right\}\right\}$

Out[33]= $3200\, m$

There are two places that the trains are in the same location at the same time. The first solution is the trivial given condition. The second line of code in Ex. 4.16 is picking out the second solution for t.

A plot of the motion of the trains is given in Figure 4.6.

Example 4.17 An object is dropped from a height of 50 meters at the same time a second object is launched upward from zero height at a speed of 30 m/s. At what time do they have the same height and what is that height?

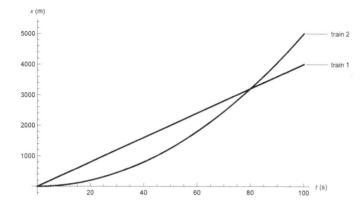

Figure 4.6 Two trains have the same position at $t = 0$. Train 1 moves at a constant speed of 40 m/s and train 2 accelerates from rest at 1 m/s^2.

In[34]:= s = Solve$\left[50 \text{ m} - \frac{1}{2} g t^2 == 30 \frac{\text{m}}{\text{s}} t - \frac{1}{2} g t^2, \{t\}\right]$

t′ = t /. s[[1]];

$50. \text{ m} - \frac{1}{2} g t′^2$

Out[34]= $\left\{\left\{t \rightarrow \frac{5}{3} \text{ s}\right\}\right\}$

Out[36]= 36.3797 m

Note that since both objects have the same acceleration, they approach each other at their constant relative speed. A plot of the motion of the objects is given in Figure 4.7.

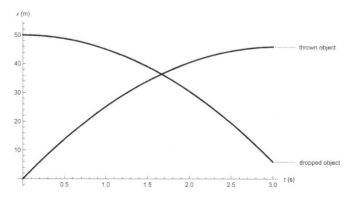

Figure 4.7 Object 1 is dropped from a height of 50 meters at the same time an object 2 is launched upward from zero height at a speed of 30 m/s.

CHAPTER 5

Motion in Two and Three Dimensions

5.1 THE VECTORS

If the position vector **r** is known, then velocity and acceleration vectors are readily computed by their definitions,

$$\mathbf{v} = \frac{d\mathbf{r}}{dt},$$

and

$$\mathbf{a} = \frac{d\mathbf{v}}{dt}.$$

If instead, the velocity was known, the position is determined from

$$\mathbf{r} = \int dt \, \mathbf{v}.$$

In a third case where the acceleration was known, the velocity is computed from

$$\mathbf{v} = \int dt \, \mathbf{a}.$$

The above equations are definitions, and thus, are always true.

5.2 PROJECTILE MOTION

Projectile motion is a straightforward extension of free-fall motion described in Sect. 4.5.3. The acceleration of vector is written

$$\mathbf{a} = -g \, \hat{\mathbf{y}},$$

DOI: 10.1201/9781003481980-5

and the velocity vector is

$$\mathbf{v} = \mathbf{v}_0 - g\,\hat{\mathbf{y}}.$$

The velocity vector is written

$$\mathbf{v} = v_x\,\hat{\mathbf{x}} + v_y\,\hat{\mathbf{y}},$$

and its solution is

$$v_x = v_{0x}$$

and

$$v_y = v_{0y} - gt.$$

The position vector is written

$$\mathbf{r} = x\,\hat{\mathbf{x}} + y\,\hat{\mathbf{y}},$$

and the solution is

$$x = x_0 + v_{0x}t$$

and

$$y = y_0 + v_{0y}t - \frac{1}{2}gt^2.$$

The resulting x-component of motion is unchanged from the initial condition and the y-component of motion is the same as free-fall.

Example 5.1 An object is launched from the ground with initial velocity vector $\mathbf{v} = (4\text{ m/s})\,\hat{\mathbf{x}} + (3\text{ m/s})\,\hat{\mathbf{y}}$. Calculate the time for the object to get to its maximum height, the maximum height, and the horizontal distance traveled when the object returns to the ground.

In[75]:= x_θ = 0.; y_θ = 0.; $v_{\theta x}$ = 4. $\dfrac{m}{s}$; $v_{\theta y}$ = 3. $\dfrac{m}{s}$;

sol = Solve$\left[0 == v_{\theta y} - g\ t,\ t\right]$; t_{top} = t /. sol$[\![1]\!]$

$v_{\theta y}\ t_{top} - \dfrac{1}{2}\ g\ t_{top}{}^2$

$v_{\theta x}\ (2\ t_{top})$

Out[76]= 0.305915 s

Out[77]= 0.458872 m

Out[78]= 1.83549 m

AI plot of the trajectory (y vs. x) for Ex. 5.1 is shown in Figure 5.1.

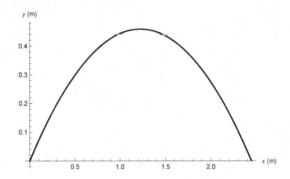

Figure 5.1 The trajectory for the projectile motion of Ex. 5.1 is a parabola.

To calculate how far the object traveled (d), we could calculate its speed (v),

$$v = \sqrt{v_{0x}^2 + v_{0y}^2} = \sqrt{v_{0x}^2 + (v_{0y} - gt)^2},$$

and integrate it over the time T that the object is in flight,

$$d = \int_0^T dt\, v = \int_0^T dt\, \sqrt{v_{0x}^2 + (v_{0y} - gt)^2},$$

Example 5.2 Calculate the total distance the object travels in Ex. 5.1.

In[85]:= $\mathtt{Integrate}\left[\sqrt{\mathtt{v_{0x}}^2 + (\mathtt{v_{0x}} - \mathtt{g\ t})^2}, \{\mathtt{t}, \mathtt{0\ s}, \mathtt{2\ t_{top}}\}\right]$

Out[85]= 2.10676 m

The total distance traveled is the arc length of the curve. Mathematica can calculate this directly with the function ArcLength[$\{x_1, ..., x_n\}, \{t, t_{min}, t_{max}\}$] which gives the length of the parametrized curve whose Cartesian coordinates x_i are functions of t.

Example 5.3 Calculate the arc length for the parabolic projectile motion trajectory.

In[73]:= $\mathtt{ArcLength}\left[\left\{\mathtt{v_{0x}\ t}, \mathtt{v_{0y}\ t} - \frac{1}{2}\,\mathtt{g\ t^2}\right\}, \{\mathtt{t}, \mathtt{0\ s}, \mathtt{2\ t_{top}}\}\right]$

Out[73]= 2.10676 m

Example 5.4 An object is launched with a speed of 10 m/s at a height of 15 m from the edge of a cliff. How long does it take to hit the ground?

In[13]:= y = -15 m; v₀ᵧ = 10 m/s; Solve$\left[y = v_{0y} t - \frac{g t^2}{2.}, t\right]$

Out[13]= $\{\{t \to -1.00488 \text{ s}\}, \{t \to 3.04431 \text{ s}\}\}$

The first of the mathematical solutions is unphysical. It corresponds to the time when the projectile viewed at negative times would have been at the height of −15 m. It is a projection of the parabola backward in time.

At the top of the trajectory,

$$v_y = 0 \text{ m/s}.$$

Example 5.5 Calculate the time it takes for the projectile of Ex. 5.6 to reach maximum height.

In[15]:= v₀ᵧ = 10 m/s; sol = Solve$\left[0. \frac{m}{s} = v_{0y} - g t, t\right]$

Out[15]= $\{\{t \to 1.01972 \text{ s}\}\}$

Example 5.6 How high does the projectile go?

In[16]:= t_top = t /. sol⟦1⟧; y = v₀ᵧ t_top $- \frac{g t_{top}{}^2}{2.}$

Out[16]= 5.09858 m

The trajectory, a plot of y vs. x, for the projectile of Ex. 5.6 is shown in Figure 5.2.

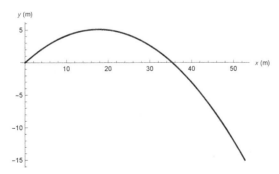

Figure 5.2   The trajectory is shown for the projectile motion of Ex. 5.6.

5.2.1 Range

The range R of a projectile is the horizontal (x) distance traveled. It is given by

$$R = v_{0x}t,$$

where t is the time that the object is in flight. For an object launched at an angle θ with a speed v_0,

$$x = v_0 \cos\theta \, t,$$

and

$$y = v_0 \sin\theta \, t - \frac{1}{2}gt^2.$$

The equation for y gives the flight time (the time to reach $y = 0$) to be

$$t = \frac{2v_0 \sin\theta}{g}.$$

The equation for x gives the range to be

$$R = \frac{2v_0^2 \cos\theta \sin\theta}{g}.$$

The maximum range can be found by setting

$$\frac{dR}{d\theta} = 0,$$

and solving for θ.

Example 5.7 Find the angle that gives the maximum range.

In[3]:= `Solve`$\left[\text{`D[Sin[`}\theta\text{`] Cos[`}\theta\text{`], `}\theta\text{`]` == 0 \&\& 0 < }\theta\text{ < }\frac{\pi}{2}\text{, }\theta\right]$ `// Simplify`

Out[3]= $\left\{\left\{\theta \to \frac{\pi}{4}\right\}\right\}$

The maximum range occurs when $\theta = \frac{\pi}{4}$.
Since

$$\sin\theta\cos\theta = \sin\left(\frac{\pi}{2} - \theta\right)\cos\left(\frac{\pi}{2} - \theta\right),$$

launching a projectile at an angle θ gives the same range as launching an angle $\frac{\pi}{2} - \theta$.

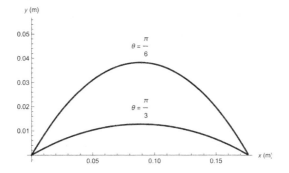

Figure 5.3 The trajectory for a projectile launched at an angle $\theta = \pi/6$ and the same range as that launched at angle $\theta/3$.

Example 5.8 Verify that $\sin\theta\cos\theta = \sin\left(\frac{\pi}{2} - \theta\right)\cos\left(\frac{\pi}{2} - \theta\right)$.

In[5]:= $\textbf{Sin[\theta] Cos[\theta]} == \textbf{Sin}\left[\frac{\pi}{2} - \theta\right]\textbf{Cos}\left[\frac{\pi}{2} - \theta\right]$

Out[5]= **True**

The trajectories for projectiles launched with the same speed but at angles of $\pi/6$ and $\pi/3$ are shown in Figure 5.4.

5.2.2 The Monkey Problem

A hunter aims his tranquilizer dart at a monkey hanging from a branch of a tree. The monkey is at a horizontal distance d and height h relative to where

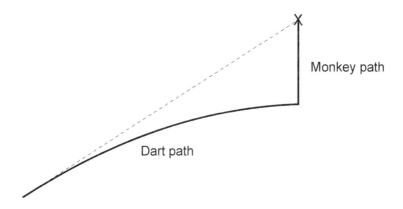

Figure 5.4 The trajectory for the dart drops the same amount per time from the dashed line as the monkey.

the dart is launched. Let the x and y speeds of the dart be v_{x0} and v_{y0}. Aiming the dart at the monkey means that

$$\frac{v_{0x}}{v_{0y}} = \frac{d}{h}.$$

At the instant the dart is fired, the monkey lets go of the branch. The resulting motion is such that the monkey and dart are both accelerated downward at g and the dart strikes the monkey at exactly the location of the aim. The monkey's vertical position y_m is

$$y_m = h - \frac{1}{2}gt^2.$$

The dart has a vertical position y_d,

$$y_d = v_{0y}t - \frac{1}{2}gt^2.$$

The paths cross when $y_m = y_d$ and this occurs at time $t = h/v_{0y}$. The horizontal position of the dart x_d is

$$x_d = v_{0x}t = \frac{v_{0x}h}{v_{0y}} = d,$$

where the last equality is true because the dart is aimed at the monkey. Therefore the monkey and the dart have the same x and y positions at the same time. The paths taken by the dart and the monkey are shown in Figure 5.4.

Example 5.9 Solve for the conditions under which the monkey and the dart have the same (x, y) coordinates at the same time.

In[6]:= $\mathbf{Solve}\left[h - \frac{1}{2} g\ t^2 == v_{\theta y}\ t - \frac{1}{2} g\ t^2\ \&\&\ v_{\theta x}\ t == d\ \&\&\ \frac{v_{\theta x}}{v_{\theta y}} == \frac{d}{h},\right.$

$\left\{t,\ v_{\theta x}\right\}\right]$

Out[6]= $\left\{\left\{t \to \frac{h}{v_{\theta y}},\ v_{\theta x} \to \frac{d\ v_{\theta y}}{h}\right\}\right\}$

Example 5.10 The height of the monkey is $h = 10$ m. The horizontal distance is $d = 10$ m. The vertical speed of the dart is $v_{0x} = 20$ m/s. Calculate the height of the monkey (y_m) when it gets hit by the dart.

In[11]:= **h = 10 m; d = 10 m; v₀ y = 20. $\dfrac{m}{s}$; v₀ x = $\dfrac{d\ v₀\ y}{h}$;**

$$t = \frac{h}{v_{0\ y}};$$

$$\textbf{UnitConvert}\left[h - \frac{1}{2}\ g\ t^2\right]$$

Out[13]= 8.77417 m

5.3   CHARGED PARTICLE IN A MAGNETIC FIELD

The motion of particle with charge q in a uniform magnetic field (B) is uniform circular motion if the velocity vector is perpendicular to the field direction. The radius of the circle is

$$r = \frac{mv}{qB}.$$

The period of the motion is

$$T = \frac{2\pi r}{v} = \frac{2\pi m}{qB}.$$

If there is a component of motion parallel to the magnetic field direction, then the motion is uniform circular superimposed with linear (Figure 5.5). The components of velocity perpendicular and parallel to the field are

$$v_\perp = v\sin\theta,$$

and

$$v_\parallel = v\cos\theta,$$

where θ is the angle between **v** and **B**. The pitch of the helix (p) is the distance between adjacent turns,

$$p = v_\parallel T.$$

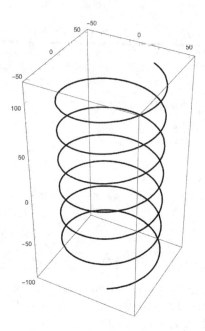

Figure 5.5 The trajectory for a charged particle moving in a magnetic field is a helix.

The pitch of a helix is the height of one complete turn.

Example 5.11 Calculate the pitch of the helix for an electron with a speed of 10^8 m/s moving at an angle of 0.1 radians w.r.t. a 0.1 T magnetic field.

In[15]:= m = [**electron** PARTICLE] [*mass*] ; v = 10⁸ $\frac{m}{s}$;

$$T = \frac{2 \pi \, m}{e \, B} ; B = .1 \, T;$$

θ = 0.1;
UnitConvert[v Cos[θ] T]

Out[18]= 0.0355454 m

CHAPTER 6

Newton's Laws

6.1 FORCES

A force is a push or a pull. Force has a direction and is a vector. There may be multiple forces acting on an object. The unit of force is the newton (N). In base units,

$1\text{N} = 1\text{kg} \cdot \text{m/s}^2.$

Example 6.1 Get the newton unit.

```
In[1]:= Quantity["newton"]
        UnitConvert[%]
```

Out[1]= 1 N

The vector sum of all the forces acting on an object is called the net force ($\mathbf{F}_{net}$).

6.2 FIRST LAW

If $\mathbf{F}_{net} = 0$, then the motion of an object is unchanged. An object at rest remains at rest and an object moving in a straight line continues to move in the straight line. The acceleration is zero,

$$\mathbf{a} = 0,$$

and the speed is constant.

Figure 6.1 shows the forces on a hockey puck that is sliding on ice at constant velocity. It will continue that motion in a straight line until disrupted by another force.

DOI: 10.1201/9781003481980-6

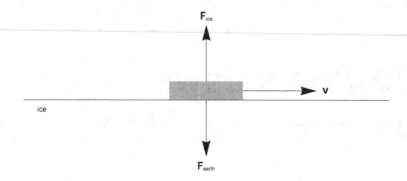

Figure 6.1 For an object sliding on ice at constant velocity, there are two forces, the downward pull of the earth and the upward push of the ice, which make zero net force.

6.3 SECOND LAW

The acceleration (a) of an object is given by

$$\mathbf{a} = \frac{\mathbf{F}_{net}}{m},$$

6.4 THIRD LAW

The third law states that if object 1 exerts a force on object 2, then object 2 must also exert a force on object 1 with the same magnitude but opposite direction. Thus, all forces occur in pairs. It is important to note that the two forces that make up a 3rd-law pair act on different objects.

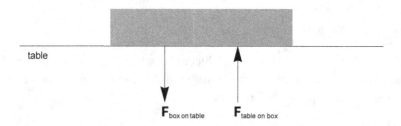

Figure 6.2 For a box sitting on a table, the force of the table pushing on the box is equal in magnitude to the force of the box pushing on the table. The forces are in opposite directions.

6.5 FREE-BODY DIAGRAMS

The free-body diagram (FBD) is extremely useful for visualizing the force vectors and serves as an essential aid for writing the second law. The easiest way to make an FBD is to place a dot at the location of the object that is being analyzed, and then draw the force vectors. One of the forces is always the force of gravity. It is "action at a distance." The universal force of gravity is covered in Chap. 10. For the other forces, there must be some physical contact. Note that $m\mathbf{a}$ is not part of the FBD; it is not a force, but rather the consequence of the net force.

To apply Newton's second law, one needs to choose a coordinate system, one uses the FBD to set the net force in each direction (x, y, and z) to be equal to the acceleration in that direction.

6.5.1 Free-Fall

The FBD for free-fall is extremely simple. There is just one force, that of gravity pointing downward as shown in Figure 6.3. The force of gravity is

$$F_g = mg.$$

Newton's second law reads

$$F_g = mg = ma.$$

This gives the magnitude of the acceleration to be

$$a = g.$$

The direction is downward.

Example 6.2 Calculate the force on a mass of 1 kg in free-fall.

```
In[2]:= θ = 30 Degree; W = UnitConvert[F Δx Cos[θ], J]
        N[W, 3]

Out[2]= 25 √3 J

Out[3]= 43.3 J
```

$\mathbf{F}_g$

Figure 6.3 The free-body diagram for an object in free-fall near the earth has just one force, the downward force of gravity.

6.5.2 Tension

Tension is a force. Tension is put on a rope by applying a force on each end. The tension exists everywhere along the length of the rope. Consider an object hanging from a rope that is attached to the ceiling (Fig 6.4). There is a tension force **T** pulling downward on the ceiling (the third-law pair to the ceiling pulling on the rope) and a tension force with the same magnitude pulling upward on the object (the third-law pair to the object pulling on the rope).

The free-body diagram for the object hanging from a rope is shown in Figure 6.5a).

Suppose the hanging object of Fig 6.4 was inside an automobile that was accelerating. The object now hangs at an angle θ as measured from the ver-

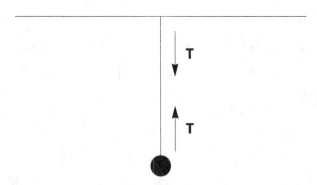

Figure 6.4 An object hangs from a rope that is attached to the ceiling. This causes tension in the rope which pulls downward on the ceiling and upward on the object.

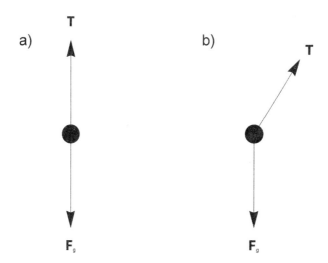

Figure 6.5 The free-body diagram for an object hanging from a string is shown for two cases: (a) the object has zero acceleration, and (b) the object accelerates to the right.

tical. The free-body diagram is shown in Figure 6.5(b). The horizontal component of **T** is causing the acceleration of the hanging object and the vertical component of **T** is balancing the force of gravity,

$$T \sin\theta = ma,$$

and

$$T \cos\theta - mg = 0,$$

where θ is the angle between the vertical and the direction of **T**.

Example 6.3 An object of mass m hangs from a string inside an automobile that has acceleration a. Calculate the relationship between the acceleration and the angle at which the objects hang and the tension in the string.

```
In[4]:= ClearAll["Global`*"];
        Fg = m g {0, -1};
        𝒯 = T {Sin[θ], Cos[θ]};
        Solve[Fg + 𝒯 == m a {1, 0}, {a, T}]

Out[5]= {{a → g Tan[θ], T → g m Sec[θ]}}
```

Note that in Ex. 6.3, T has been used for the magnitude of the tension and a different variable $\mathcal{T}$ was used for the tension vector.

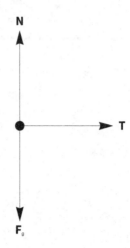

Figure 6.6 An object slides along a horizontal surface without friction as it is being pulled via an attached string. The free-body diagram has three forces.

Figure 6.6 shows the FBD of an object that is sliding on a horizontal surface without friction as it is being pulled by a string. Newton's second law is

$$T = ma,$$

and

$$N - mg = 0.$$

6.5.3 Pulley

A pulley may be used to change the direction of a tension force. Consider a mass M sliding on a horizontal surface while attached to mass m by a string that passes through a pulley (Figure 6.7).

For mass M, the 2nd law reads

$$T = ma,$$

and

$$N - mg = 0.$$

For mass m, the 2nd law reads

$$mg - T = ma.$$

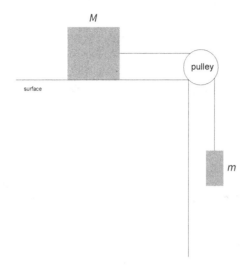

Figure 6.7 The mass M is connected to the mass m by a string that passes through a pulley.

Example 6.4 Solve for the acceleration and tension.

```
In[6]:= ClearAll["Global`*"];
        Solve[T == M a && m g - T == m a, {T, a}]
```
$$Out[6]= \left\{\left\{T \rightarrow \frac{g\,m\,M}{m + M}, \; a \rightarrow \frac{g\,m}{m + M}\right\}\right\}$$

6.6 FRICTION

There are two types of friction. Static friction acts to prevent motion from an applied force. Consider a force **F**, applied to a stationary object as shown in Figure 6.8. The static friction ($\mathbf{f_s}$) exactly cancels the applied force up to the point where the object starts to slide,

$$\mathbf{f_s} = \mathbf{F}.$$

The maximum force of static friction (f_s^{max}) is proportional to the normal force and is written in terms of a dimensionless constant μ_s,

$$f_s^{max} = \mu_s N.$$

The direction of the force of static friction is opposite the direction in which the object would slide if friction was absent.

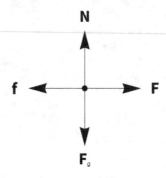

Figure 6.8 A force **F** is applied to a stationary object. The force **f** is static friction if the object does not move and kinetic friction if it slides.

It is important to note that $\mu_s N$ is the maximum value of static friction and that, in general, the static friction could be smaller than this value.

If the applied force is greater than f_s^{max}, then the situation is no longer static and the object moves. Static friction no longer applies. A moving object has a force of kinetic friction acting in a direction opposite to the motion. The force of kinetic friction (f_k) is written as

$$f_k = \mu_k N,$$

where μ_k is the coefficient of kinetic friction. It is readily noticed that force needed to overcome static friction to start an object sliding is greater than that needed to keep it sliding, so $\mu_k < \mu_s$.

The force of friction is summarized in Figure 6.9 which shows the frictional force f as a function of the applied force F. The force of static friction will increase and exactly balance the applied force until the force of static friction reaches its maximum. At that point the object moves and the frictional force is kinetic (and constant).

6.6.1 Pulley Problem

Consider the pulley problem of Figure 6.7 and add kinetic friction to the sliding mass M. The free-body diagram for the sliding mass (Figure 6.6) now has an additional force of kinetic friction that opposes the tension.

Example 6.5 Solve for the acceleration and tension.

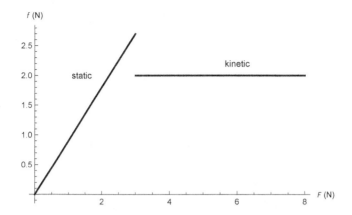

Figure 6.9 A force F is applied to a stationary object. The object does not move until the force exceeds the maximum force of static friction.

In[7]:= **ClearAll["Global`*"];**
Solve[T - f$_k$ == M a && m g - T == m a, {T, a}]

Out[7]= $\left\{\left\{T \rightarrow -\dfrac{-g\,m\,M - m\,f_k}{m + M}\,,\ a \rightarrow -\dfrac{-g\,m + f_k}{m + M}\right\}\right\}$

6.6.2 Mass Pinned Against a Wall

Consider a mass m that is pinned against a vertical wall by a force $\mathbf{F}$ that acts at an angle θ w.r.t. the horizontal (see Figure 6.10). The FBD is shown in Figure 6.11.

Example 6.6 Solve for the minimum force needed to keep the mass from sliding.

In[8]:= **ClearAll["Global`*"];**
Solve[F Cos[θ] == N && f$_s$ + F Sin[θ] == m g && f$_s$ == μ_s N,
{F, N, f$_s$}]

Out[8]= $\left\{\left\{F \rightarrow \dfrac{g\,m}{Sin[\theta] + Cos[\theta]\,\mu_s}\,,\right.\right.$
$\left.\left. N \rightarrow \dfrac{g\,m\,Cos[\theta]}{Sin[\theta] + Cos[\theta]\,\mu_s}\,,\ f_s \rightarrow \dfrac{g\,m\,Cos[\theta]\,\mu_s}{Sin[\theta] + Cos[\theta]\,\mu_s}\right\}\right\}$

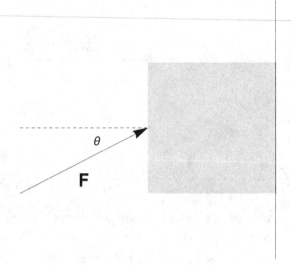

Figure 6.10 A force F is applied at an angle θ to a mass m to keep it from sliding vertically along the wall.

6.7 BLOCK ON INCLINED PLANE

Consider a block on a plane that has an inclination angle θ as shown in Figure 6.12.

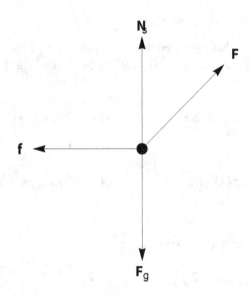

Figure 6.11 The free-body diagram for the mass pinned against the wall has four forces.

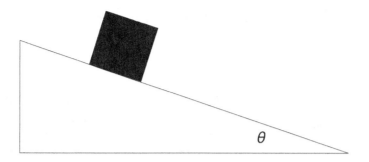

Figure 6.12   A block is on a plane that has an inclination angle θ.

6.7.1   No Friction

For the case of zero friction, there are just two forces, gravity and the normal force. The free-body diagram is shown in Figure 6.13.

Figure 6.13   Free-body diagram for a block on an inclined plane with no friction.

If we choose the unit vectors to be horizontal and vertical, then

$$\mathbf{F}_g = -mg\,\hat{\mathbf{y}},$$

and

$$\mathbf{N} = N\sin\theta\,\hat{\mathbf{x}} + N\cos\theta\,\hat{\mathbf{y}}.$$

The acceleration is along the inclined plane and may be written as

$$a = a\cos\theta\,\hat{\mathbf{x}} - a\sin\theta\,\hat{\mathbf{y}}.$$

Example 6.7 Calculate the normal force and the acceleration for the block on an inclined plane.

```
In[9]:= $Assumptions = {θ > 0, θ < π/2};
       Fg = m g {0, -1}; N = N {Sin[θ], Cos[θ]};
       Solve[Fg + N == m a {Cos[θ], -Sin[θ]}, {N, a}] //
       Simplify
```

```
Out[10]= {{N → g m Cos[θ], a → g Sin[θ]}}
```

Since we know the direction of the acceleration, we might as well choose that direction as $\hat{\mathbf{x}}$. Then

$$\mathbf{F}_g = mg\sin\theta\,\hat{\mathbf{x}} - mg\cos\theta\,\hat{\mathbf{y}},$$

$$\mathbf{N} = N\,\hat{\mathbf{y}},$$

and

$$\mathbf{a} = a\,\hat{\mathbf{x}}.$$

Example 6.8 Calculate the normal force and the acceleration for the block on an inclined plane using coordinates with $\hat{\mathbf{x}}$ along the inclined plane.

```
In[11]:= $Assumptions = {θ > 0, θ < π/2};
       Fg = m g {Sin[θ], -Cos[θ]}; N = N {0, 1};
       Solve[Fg + N == m a {1, 0}, {N, a}] // Simplify
```

```
Out[12]= {{N → g m Cos[θ], a → g Sin[θ]}}
```

Now add a horizontal force **F**. Note that this increases the normal force because **F** has a component perpendicular to **N**. The free-body diagram is shown in Figure 6.14.

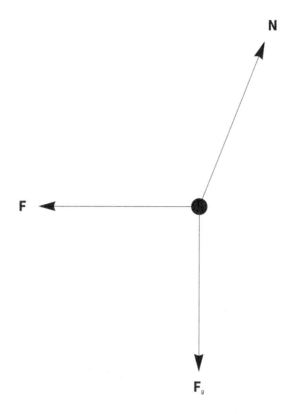

Figure 6.14 Free-body diagram for a block on an inclined plane with an applied horizontal force **F**.

Example 6.9 Find the value of the horizontal force and the normal force if the block has zero acceleration.

```
In[13]:= $Assumptions = {θ > 0, θ < π / 2};
       Fg = m g {Sin[θ], -Cos[θ]}; N = N {0, 1};
       𝓕 = F {-Cos[θ], -Sin[θ]};
       Solve[Fg + N + 𝓕 == 0, {F, N}]
```

Out[13]= {{F → g m Tan[θ], N → g m Cos[θ] + g m Sin[θ] Tan[θ]}}

6.7.2 Static Friction

For the case of static friction, the free-body diagram is shown in Figure 6.15.

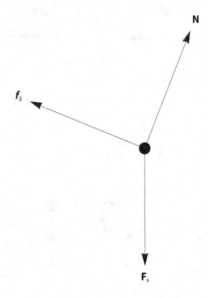

Figure 6.15 Free-body diagram for a block on an inclined plane with static friction.

Example 6.10 Calculate the normal force and the friction force for the block on an inclined plane with static friction.

In[14]:= **$Assumptions = {θ > 0, θ < π/2};**
Fg = m g {Sin[θ], -Cos[θ]}; N = N {0, 1};
f = f {-1, 0}; Solve[Fg + N + f == 0, {f, N}] // Simplify

Out[14]= **{{f → g m Sin[θ], N → g m Cos[θ]}}**

Example 6.11 Calculate the relationship between μ_s and the inclination angle θ when the force of static friction is at its maximum.

In[15]:= **$Assumptions = {θ > 0, θ < π/2};**
Fg = m g {Sin[θ], -Cos[θ]}; N = N {0, 1};
f = μ_s N {-1, 0};
Solve[Fg + N + f == 0, {μ_s, N}] // Simplify

Out[15]= **{{μ_s → Tan[θ], N → g m Cos[θ]}}**

6.7.3 Kinetic Friction

If the block slides, then the frictional force is kinetic.

Example 6.12 Calculate the normal force and the acceleration of the block on an inclined plane with kinetic friction.

In[16]:= $Assumptions = {θ > 0, θ < π/2, N < mg, m > 0, g > 0};
        F_g = m g {Sin[θ], -Cos[θ]}; N = N {0, 1};
        f = μ_k N {-1, 0};
        Solve[F_g + N + f == m a {1, 0}, {a, N}] // Simplify

Out[16]= {{a → g (Sin[θ] - Cos[θ] μ_k), N → g m Cos[θ]}}

6.8   UNIFORM CIRCULAR MOTION

Uniform circular motion means constant speed v with the velocity vector tangent to the circle (see Figure 6.16). The velocity vector is not constant because its direction is changing. The angular variable (θ) is the ratio of arc length to radius (r) of the circle,

$$\theta = \frac{vt}{r}.$$

Its time derivative is

$$\frac{d\theta}{dt} = \frac{v}{r}.$$

The velocity vector is

$$\mathbf{v} = -v\sin\theta\,\hat{\mathbf{x}} + v\cos\theta\,\hat{\mathbf{y}}.$$

Taking the derivative,

$$\mathbf{a} = \frac{d\mathbf{v}}{dt} = \frac{d\mathbf{v}}{d\theta}\frac{d\theta}{dt} = (-v\cos\theta\,\hat{\mathbf{x}} - v\sin\theta\,\hat{\mathbf{y}})\frac{v}{r} = -\frac{v^2}{r}.(\cos\theta\,\hat{\mathbf{x}} + \sin\theta\,\hat{\mathbf{y}})$$

The acceleration vector points toward the center of the circle and its magnitude is

$$a = \frac{v^2}{r}.$$

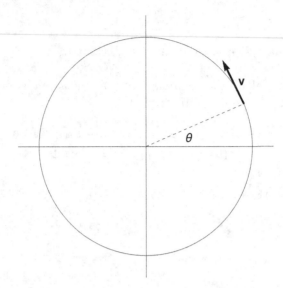

Figure 6.16 An object with uniform circular motion has a uniform speed and a velocity vector that is tangent to the circle.

Example 6.13 Calculate the acceleration vector for uniform circular motion.

```
In[17]:= $Assumptions = {v > 0, r > 0};
        v = v {Cos[θ[t]], -Sin[θ[t]]}; a = D[v, t] /. θ'[t] → v/r
```

$$Out[18]= \left\{ -\frac{v^2 \sin[\theta[t]]}{r}, -\frac{v^2 \cos[\theta[t]]}{r} \right\}$$

6.8.1 Object Connected to a String

Consider an object connected to a string of length L (Figure 6.17) undergoing uniform circular motion with speed v and radius r. The string makes an angle θ with the vertical. The orbit radius is

$$r = L\sin\theta.$$

The horizontal component of the tension is the centripetal force and the vertical component balances the force of gravity.

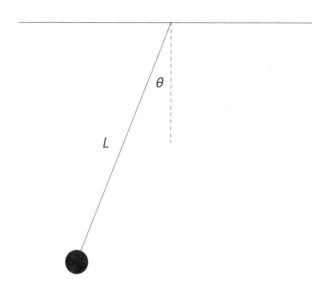

Figure 6.17 An object hangs from a string at an angle θ as it undergoes uniform circular motion.

Example 6.14 Calculate the speed of the object and the tension in the string.

```
In[19]:= $Assumptions = {θ > 0, θ < π/2, v > 0, g > 0, m > 0, r > 0};
        Fg = m g {0, -1}; 𝒯 = T {Sin[θ], Cos[θ]}; r = L Sin[θ];
        Solve⎡Fg + 𝒯 == m (v²/r) {1, 0}, {v, T}⎤ // Simplify

Out[19]= {{v → √(g L Sin[θ] Tan[θ]), T → g m Sec[θ]}}
```

6.8.2   Car Rounding a Curve

For a car rounding a curve on a level road, the force of static friction, which prevents the car from sliding, causes the centripetal acceleration. The FBD is shown in Figure 6.19.

Example 6.15 Calculate the maximum speed that a car can have without sliding when rounding a curve of radius r when the coefficient of static fiction is μ_s.

Figure 6.18 Free-body diagram for uniform circular motion of an object hanging from a string.

In[20]:= **Clear[r];**
$Assumptions = {θ > 0, θ < π/2, v > 0, g > 0, m > 0,
r > 0, μ$_s$ > 0};
F$_g$ = m g {0, -1}; N = N {0, 1}; f = μ$_s$ N {1, 0};

$$\text{Solve}\left[\text{F}_g + N + f == m\,\frac{v^2}{r}\,\{1, 0\}, \{v, N\}\right] \text{// Simplify}$$

Out[21]= $\left\{\left\{v \to \sqrt{g\,r\,\mu_s}\,,\; N \to g\,m\right\}\right\}$

It is common practice to bank a curve on a road, which gives the road a slope such that if the car slides due to inadequate friction, the sliding would be uphill. The normal force has a component that points toward the center of the circle and provides the centripetal force needed to round the curve without slipping. The free-body diagram is shown in Figure 6.20. The banking angle is the angle that the normal force vector makes with the vertical ($\hat{y}$) direction.

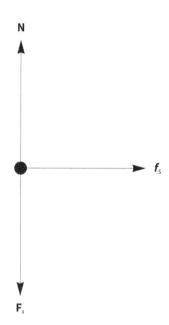

Figure 6.19 Free-body diagram for a car rounding a curve. The centripetal force is that of static friction.

Example 6.16 Suppose the static friction on a section of road was zero. Calculate the maximum speed that a car can have without sliding perpendicular to its motion when rounding a curve of radius r when the banking angle is θ.

In[22]:= $Assumptions = {θ > 0, θ < π / 2, v > 0, g > 0, m > 0, r > 0};
Fg = m g {0, -1}; N = N {Sin[θ], Cos[θ]};

Solve[Fg + N == m $\frac{v^2}{r}$ {1, 0}, {v, N}] // Simplify

Out[23]= {{v → $\sqrt{g r \, Tan[θ]}$, N → g m Sec[θ]}}

6.9   SIMPLE HARMONIC MOTION

Simple harmonic motion occurs when you have a linear restoring force, when, for example, a displacement to the right produces a force to the left in proportion to the displacement. This occurs in many scenarios, the simplest being a mass sliding on a horizontal surface connected by a spring (Figure 6.21). The position is a function of time and Newton's 2nd law is

Figure 6.20 Free-body diagram for a car rounding a banked curve. The centripetal force is caused by the horizontal component of the normal force.

written

$$F = -kx = m\frac{d^2x}{dt^2}.$$

The constant k is called the spring force and its units are N/m. The sine and cosine are the only functions that give themselves back again with a minus sign after taking two derivatives.

Example 6.17 Take two derivatives of the $\sin(x)$ and $\cos(x)$.

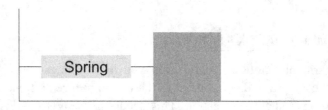

Figure 6.21 A mass slides on a frictionless surface and alternately compresses and stretches the spring.

In[24]:= $\partial_x \partial_x$ Sin[x]
$\partial_x \partial_x$ Cos[x]

Out[24]= -Sin[x]

Out[25]= -Cos[x]

One may write the position as a function of time as

$$x(t) = A\cos(\omega t),$$

where A is the amplitude, its maximum displacement, and ω is the angular frequency. Note the choice of cosine rather than sine is purely a matter of initial position at $t = 0$. The angular frequency is related to the period (T) of the motion by

$$\omega = \frac{2\pi}{T}$$

The wave equation gives the relationship between k, m, and ω.

Example 6.18 Find the relationship between k, m, and ω.

In[26]:= $Assumptions = {m > 0};
x = A Cos[ω t];
Solve[-k x == m $\partial_t \partial_t$ x, k]

Out[26]= $\left\{\left\{k \to m\,\omega^2\right\}\right\}$

Example 6.19 A spring has $k = 5\ \frac{N}{m}$. What is the period of oscillation for a 100 g mass?

In[27]:= k = 5 $\frac{N}{m}$; m = 100. g; ω = $\sqrt{\frac{k}{m}}$; UnitConvert$\left[T = \frac{2\pi}{\omega}\right]$

Out[27]= 0.888577 s

The velocity and acceleration are determined by taking derivatives,

$$v(t) = \frac{dx}{dt} = -\omega A \sin(\omega t),$$

and

$$a(t) = \frac{dv}{dt} = -\omega^2 A \cos(\omega t).$$

Example 6.20 Consider the same spring of Ex. 6.19. The mass is released from rest at a position of 1 cm. Determine the position, velocity, and acceleration at $t = 10$ s.

In[28]:= **A = 1 cm; x = A Cos[ω t]; v = ∂ₜ x; a = ∂ₜ v;**
UnitConvert[x /. t → 10 s, cm]

UnitConvert$\left[\text{v /. t → 10 s, } \dfrac{m}{s}\right]$

UnitConvert$\left[\text{a /. t → 10 s, } \dfrac{m}{s^2}\right]$

Out[29]= -0.0248409 cm

Out[30]= -0.0706889 m/s

Out[31]= 0.0124204 m/s^2

CHAPTER 7

Energy

7.1   WORK

When a force acts on an object, causing the object to move, this scenario is described with the concept of work. There may be multiple forces acting and one needs to be careful to specify which force, or combination of forces, is used to calculate the work. In the simplest case, work (W) is defined to be force (F) times magnitude of the displacement (Δx),

$$W = F\Delta x.$$

Work is a scalar quantity and has units of joules (J). This simplest form applies to the case where the force is constant and is in the same direction as the displacement (Figure 7.1).

Example 7.1 A force of 10 N pushes on an object causing it to move 5 m in the direction of the force. Calculate the work done by the force on the object.

```
In[1]:= F = 10 N; Δx = 5 m; W = UnitConvert[F Δx, J]

Out[1]= 50 J
```

If the force is applied at an angle θ (Figure 7.2), then the work is

$$W = \mathbf{F} \cdot \Delta\mathbf{x} = F\Delta x \cos\theta.$$

Example 7.2 A force of 10 N pushes on an object at an angle θ w.r.t. the direction of motion causing it to move 5 m. Calculate the work done by the force on the object.

DOI: 10.1201/9781003481980-7

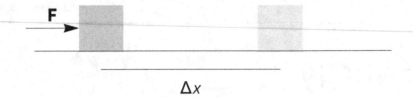

Figure 7.1 A force pushes on an object in the direction of motion.

```
In[2]:= θ = 30 Degree; W = UnitConvert[F Δx Cos[θ], J]
        N[W, 3]
```

Out[2]= $25\sqrt{3}$ J

Out[3]= 43.3 J

The normal force does no work ($\cos\theta = 0$). The downward component of the force is balanced by an increase in the normal force and also does no work.

If, in addition, the force varies as the object moves, then the work is given by

$$W = \int dx \cdot F.$$

Example 7.3 An object experiences a force, $F = (5\ N/m^2)x^2 - (7\ N/m)x$ at an angle $\theta = 20°$ w.r.t. the direction of motion. Calculate the work done by the force as the object moves from $x = 2$ m to $x = 5$ m.

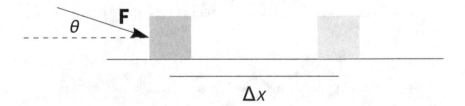

Figure 7.2 A force pushes on an object in the direction of motion.

In[4]:= **θ = 20 Degree;**

$$F = 5. \ \frac{N}{m^2} \ x^2 - 7. \ \frac{N}{m} \ x; \ W = \int_{2 \ m}^{5 \ m} F \ Cos[\theta] \ dx;$$

UnitConvert[W, J]

Out[5]= 114.173 J

Figure 7.3 shows a mass sliding down an inclined plane with friction. A force is directed up the plane in the opposite direction of the motion.

Example 7.4 A mass of 380 kg slides a distance of 3.5 m along a 27° inclined plane with a coefficient of kinetic friction of 0.2. A force **F** is directed along the plane to make the acceleration equal to zero. Calculate the work done by the force **F**, friction, gravity, and the normal force. Calculate the net work done by all the forces.

In[424]:= **d = 3.5 m; θ = 27 Degree;**

m = 380 kg;

μ_k = 0.2; N = m g Cos[θ]; f_k = μ_k N;

F = m g Sin[θ] - f_k;

W_F = UnitConvert[-F d, J]

W_f = UnitConvert[-f_k d, J]

W_g = UnitConvert[m g d Sin[θ], J]

W_N = UnitConvert[N d Cos[90 Degree], J]

W_{net} = Round[W_F + W_f + W_g + W_N, 10^{-10}]

Out[428]= -3597.08 J

Out[429]= -2324.25 J

Out[430]= 5921.33 J

Out[431]= 0. J

Out[432]= 0 J

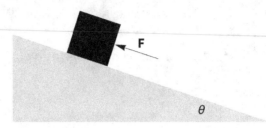

Figure 7.3 An object slides down an inclined plane with friction while a force is directed along the plane against the direction of motion.

7.2 KINETIC ENERGY

The kinetic energy(K), for non-relativistic objects, is defined to be

$$K = \frac{1}{2}mv^2.$$

Kinetic energy is a scalar quantity with units of joules. An object is non-relativistic if its speed is much smaller than the speed of light (c).

Example 7.5 Get the speed of light.

```
In[6]:= Quantity["SpeedOfLight"]
        UnitConvert[%]
```

```
Out[6]= c
```

```
Out[7]= 299 792 458 m / s
```

Example 7.6 Calculate the kinetic energy of 3 kg mass that has a speed of 14 m/s.

```
In[8]:= m = 3 kg; v = 14 m/s; K = 1/2 m v²; UnitConvert[K, J]
```

```
Out[8]= 294 J
```

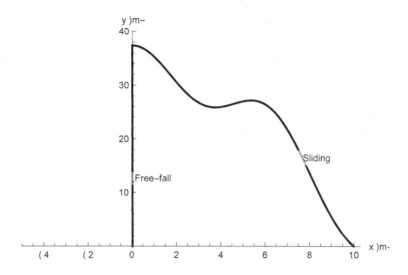

Figure 7.4 For path 1, the object is in free-fall, while for path 2, the object is sliding along a track without friction.

7.3 WORK-KINETIC ENERGY THEOREM

The work-kinetic energy theorem states that the net work done by all forces (the net force, F_{net}) is equal to the change in kinetic energy,

$$W_{net} = \int dx \cdot F_{net} = \Delta K.$$

As an example, consider an object moving without friction or air resistance along two different paths (Figure 7.4). For path 1 (free-fall), there is only one force, gravity. For path (sliding), there is also a normal force. The net work is the work done by gravity (W_g) plus the work done by the normal force (W_N). Since the normal force is always perpendicular to the motion, it does zero work, and

$$W_{net} = W_g + W_N = W_g.$$

Thus, for both scenarios, free-fall and sliding,

$$W_{net} = W_g = mgh = \Delta K = \frac{1}{2}mv^2.$$

Solving for the speed v, one gets

$$v = \sqrt{2gh}.$$

The work-kinetic energy theorem has vastly simplified the analysis of this problem.

Consider a block sliding down an inclined plane (Figure 6.12) with friction.

Example 7.7 A mass of 2 kg slides from rest down a 25° incline that has a coefficient of kinetic friction of 0.4. Calculate the kinetic energy and speed of the mass after a vertical drop of 1 m.

```
In[9]:= θ = 25 Degree; m = 2 kg; h = 1 m;
        μk = 0.4; N = m g Cos[θ];
        fk = μk N;
              h
        Δx = ─────── ;
             Sin[θ]
        K = (m g Sin[θ] - fk) Δx; UnitConvert[K, J]

                         ┌    ┌───── ┐
        v = UnitConvert  │   √ 2 K   │
                         │     ───   │
                         └      m    ┘
```

```
Out[11]= 2.78896 J
```

```
Out[12]= 1.67002 m/s
```

Example 7.8 A baseball has a mass of 0.15 kg. The baseball is hit with an initial speed of 40 m/s, and it lands at a position 15 m higher with a speed of 20 m/s. Calculate the work done by air resistance.

```
In[24]:= m = 0.15 kg; vi = 40 m/s; vf = 30 m/s; h = 15 m;
        Wg = -m g h;
              1        1
        ΔK = ─ m vf² - ─ m vi²; Wa = ΔK - Wg;
              2        2
        NumberForm[UnitConvert[Wa, J], 3]
```

```
Out[27]//NumberForm=
        -30.4 J
```

Consider the simple harmonic motion of a mass oscillating from a spring (Figure 6.21).

Example 7.9 The spring constant is 10^4 N/m. Calculate the work done by the spring as the mass moves from a displacement of −2 cm to 0 cm.

In[13]:= $k = 10^4 \dfrac{N}{m}$; $x_{max} = 2$ cm ; $W = \displaystyle\int_{-x_{max}}^{0 \, m} -k \, x \, dx$;

UnitConvert$\left[W, J\right]$

Out[14]= 2 J

7.4 POWER

Power (P) is the rate that work is done, or the rate of energy transfer,

$$P = \frac{\Delta W}{\Delta t}.$$

Horsepower (hp) is a common unit for engine power.

Example 7.10 Get the hp unit.

In[15]:= UnitConvert$\left[1. \, hp, W\right]$

Out[15]= 745.7 W

The unit kWh is common for electric energy.

Example 7.11 Convert kWh to J.

In[24]:= UnitConvert$\left[10^3 \, W \, h, J\right]$

Out[24]= 3 600 000 J

Example 7.12 An elevator has a loaded weight of 20kN. It is designed to move 25 floors (3.5 m per floor) in 20 s. Calculate the power assuming that all the work to lift the elevator comes from the motor.

In[17]:= `F = 20 kN; Δz = 25 (3.5 m); Δt = 20 s;`

$$P = \frac{F\ \Delta z}{\Delta t}; \quad \texttt{UnitConvert}[P, W]$$

`UnitConvert[P, hp]`

Out[18]= `87 500. W`

Out[19]= `117.339 hp`

Taking the limit as $\Delta t \to 0$, the power becomes

$$P = \frac{dW}{dt} = \frac{d(\mathbf{F} \cdot \mathbf{r})}{dt},$$

where $\mathbf{r}$ is the displacement vector of object that is pushed by the force $\mathbf{F}$. For a constant force in the direction of displacement,

$$P = Fv,$$

where v is the speed.

Example 7.13 A horse pulls a cart weighing 255 N at an angle of 28° w.r.t. the direction of motion at a speed of 3.3 m/s. Calculate the power output of the horse.

In[20]:= `F = 255 N; θ = 28 Degree; v = 3.3` $\frac{m}{s}$ `;`

`P = UnitConvert[F Cos[θ] v, hp]`

Out[21]= `0.99638 hp`

Consider a car that is driven by constant power. The force is

$$F = m\frac{dv}{dt} = \frac{P}{v}.$$

The power is

$$P = mv\frac{dv}{dt}.$$

Integrate over distance x,

$$\int dx\, P = \int dx\, mv\frac{dv}{dt} = \int dv\, mv^2.$$

Solving for v,

$$Px = m\frac{v^3}{3},$$

or

$$v = \left(\frac{3Px}{m}\right)^{1/3}.$$

Example 7.14 A 2000 kg car puts out a constant 200 hp over 0.25 mile. Calculate the speed obtained from rest.

```
In[22]:= P = 200 hp ; x = 0.25 mi ; m = 2000 kg ;
        v = UnitConvert[(3 P x / m)^(1/3), mi/h]
```

```
Out[23]= 100.249 mi/h
```

The power output of an airplane is proportional to v^3 in order to maintain a constant cruise speed of v,

$$P = Cv^3 = Fv.$$

Example 7.15 How much does the power have to increase the speed by 10%? By 25%?

```
In[24]:= ClearAll["Global`*"];
        v^3 /. v → 1.1
        v^3 /. v → 1.25
```

```
Out[25]= 1.331
```

```
Out[26]= 1.95313
```

7.5 POTENTIAL ENERGY

The change in kinetic energy of the falling object in Figure 7.4 did not depend on the path taken. The force of gravity is called a *conservative* force and the concept of potential energy is extremely useful.

After picking a arbitrary zero point, the potential energy (U) at a height h above that zero point is written

$$U = mgh.$$

For a conservative force, the work done in going around a closed path (ending at the starting point) is zero.

An example of a non-conservative force is friction.

7.6 ENERGY CONSERVATION

The quantity $U + K$ is conserved. Conservation of energy is written,

$$K_i + U_i = K_f + U_f,$$

where the subscripts i and f refer to the initial and final states.

The reason why energy conservation works is that the change in potential energy is defined to be the negative of the work done,

$$\Delta U = -W_{net},$$

giving

$$\Delta U + \Delta K = \Delta(U + K) = 0.$$

Therefore, $U + K$ is constant.

Example 7.16 A mass free-falls a distance of 1 m from rest. Calculate the speed.

In[38]:= K_i = 0; U_i = m g h; U_f = 0; K_f = K_i + U_i - U_f;

$$v = UnitConvert\left[\sqrt{\frac{2. \ K_f}{m}}\right] /. h \rightarrow 1 \ m$$

Out[39]= 4.42869 m/s

Consider a mass m that slides without friction on a spherical surface (Figure 7.5). At some height y, the object loses contact with the surface. This corresponds to zero normal force, and Newton's second law for circular motion reads

$$mg\cos\theta = m\frac{v^2}{r}.$$

Example 7.17 A mass slides without friction along a circular surface. At what angle does it lose contact with the surface?

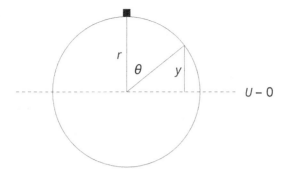

Figure 7.5 An object slides without friction on a spherical surface of radius r.

```
In[44]:= ClearAll["Global`*"];
        $Assumptions = {0 < θ < π/2, v > 0, r > 0, m > 0, g > 0};
        Uᵢ = m g r; U_f = m g y; Kᵢ = 0;
        K_f = 1/2 m v²; y = r Cos[θ];

        Solve[K_f + U_f == Kᵢ + Uᵢ && m g Cos[θ] == m v²/r, {θ, v}]
```

$$\text{Out[47]=} \; \left\{\left\{θ → ArcCos\left[\frac{2}{3}\right], \; v → \sqrt{\frac{2}{3}} \; \sqrt{g\,r}\right\}\right\}$$

Example 7.18 A mass of 1 kg slides without friction dropping a vertical height of 2 m where it compresses a spring with $k = 80$ N/m. How much does the spring compress?

```
In[5]:= h = 2 m; m = 1 kg; k = 80. N/m;

        Kᵢ = 0 J; Uᵢ = m g h + 0 J;

        K_f = 0 J; U_f = 0 J + 1/2 k x²;

        Solve[K_f + U_f == Kᵢ + Uᵢ, x]
```

$$\text{Out[8]=} \; \left\{\{x → -0.700237\,m\}, \; \{x → 0.700237\,m\}\right\}$$

Only the positive value is physical.

CHAPTER 8

Momentum

8.1 LINEAR MOMENTUM

Using Newton's second law,

$$\mathbf{F}_{\text{net}} = m\mathbf{a} = m\frac{d\mathbf{v}}{dt} = \frac{d(m\mathbf{v})}{dt},$$

we define the momentum $\mathbf{p}$ as

$$\mathbf{p} = m\mathbf{v},$$

to get

$$\mathbf{F}_{\text{net}} = \frac{d\mathbf{p}}{dt}.$$

When two objects push or pull on each other, they form a Newton's 3rd law pair whose forces cancel, and any net force must be external. If there is no external force, then

$$\frac{d\mathbf{p}}{dt} = 0,$$

and momentum cannot change. In this case the initial ($\mathbf{p}_i$) and final ($\mathbf{p}_f$) momentums are equal,

$$\mathbf{p}_i = \mathbf{p}_f.$$

8.2 IMPULSE

The impulse ($\mathbf{I}$) is defined to be the time-integral of the force,

$$\mathbf{I} = \int dt\, \mathbf{F}.$$

DOI: 10.1201/9781003481980-8

Example 8.1 A tennis ball of mass 59 g is hit from rest at a speed of 100 mi/h with a contact time of 5 ms. What is the force on the tennis ball?

In[1]:= m = 59. g; v = 100 $\frac{mi}{h}$; Δt = 5 ms; UnitConvert$\left[\frac{m\ v}{\Delta t}, N\right]$

Out[1]= 527.507 N

8.3 MOMENTUM CONSERVATION

8.3.1 Block Sliding on a Wedge

Consider a block that slides on the frictionless surface of a wedge (Figure 8.1). The wedge is free to slide on a horizontal frictionless surface, so as the block slides to the right, the wedge slides to the left. Let the velocity of the block relative to the wedge be

$$\mathbf{v} = v\cos\theta\,\hat{\mathbf{x}} - v\sin\theta\,\hat{\mathbf{y}}.$$

Let the velocity of the wedge relative to the horizontal surface be

$$\mathbf{V} = -V\,\hat{\mathbf{x}}.$$

After the block falls to a height of h, the horizontal speed of the block relative to the horizontal surface is $v\cos\theta + V$, a speed that is smaller than v. Conservation of energy gives

$$\frac{1}{2}MV^2 + \frac{1}{2}m(v\cos\theta + V)^2 + \frac{1}{2}m(v\sin\theta)^2 = mgh.$$

Conservation of the horizontal component of momentum gives

$$MV = m(v\cos\theta + V).$$

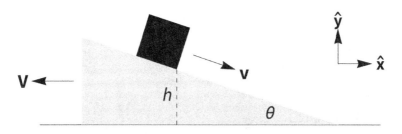

Figure 8.1 A block slides on the frictionless surface of a wedge while the wedge is free to slide along a frictionless horizontal surface.

Example 8.2 Find the speeds of the block and the wedge.

```
In[2]:= ClearAll["Global`*"];
```

$$\$Assumptions = \left\{V < 0, \; v > 0, \; 0 < \alpha < \frac{\pi}{2}, \; g > 0, \; h > 0,\right.$$

$$\left. M > 0, \; m > 0\right\};$$

$$\texttt{Solve}\left[M\,V + m\,(v\,Cos[\alpha] + V) == 0 \;\&\&\right.$$

$$\frac{1}{2}\,M\,V^2 + \frac{1}{2}\,m\,(v\,Cos[\alpha] + V)^2 + \frac{1}{2}\,m\,(v\,Sin[\alpha])^2 == m\,g\,h,$$

$$\left.\{V, \; v\}\right] \; // \; \texttt{Simplify}$$

$$Out[2]= \left\{\left\{V \rightarrow -2\,m\,Cos[\alpha]\,\sqrt{-\frac{g\,h}{(m+M)\,(-m-2\,M+m\,Cos[2\,\alpha])}}\,,\right.\right.$$

$$\left.\left.v \rightarrow 2\,(m+M)\,\sqrt{-\frac{g\,h}{(m+M)\,(-m-2\,M+m\,Cos[2\,\alpha])}}\right\}\right\}$$

Example 8.3 Give numerical answers to Ex. 8.2 for $h = 1$ m, $\theta = \pi/6$, $M = 1$ kg, and $m = 0.1$ kg.

```
In[3]:= UnitConvert[
```

$$V \; /. \; \%[[1]] \; /. \; \left\{h \rightarrow 1 \; m, \; \theta \rightarrow \frac{\pi}{6}, \; M \rightarrow 1 \; kg, \; m \rightarrow 0.1 \; kg,\right.$$

$$\left. g \rightarrow g\right\}\right]$$

$$\texttt{UnitConvert}\left[\right.$$

$$v \; /. \; \%\%[[1]] \; /. \; \left\{h \rightarrow 1 \; m, \; \theta \rightarrow \frac{\pi}{6}, \; M \rightarrow 1 \; kg, \; m \rightarrow 0.1 \; kg,\right.$$

$$\left. g \rightarrow g\right\}\right]$$

$$Out[3]= -0.3612 \; m/s$$

$$Out[4]= 4.58786 \; m/s$$

8.4 COLLISIONS

There are two categories of collisions. One of them is called elastic and corresponds to kinetic energy being conserved. The other is called inelastic, in which case kinetic energy is not conserved. There is a special subset of inelastic collisions in which the objects stick together, called a completely inelastic collision.

Example 8.4 A mass of 0.4 kg moving at a speed of 20 m/s strikes a 4 kg block and sticks to it. Calculate the speed of the combined masses after the collision.

In[5]:= v = 20 $\frac{m}{s}$; m = 0.4 kg; M = 4 kg;

$$\frac{m \, v}{m + M}$$

Out[6]= 1.81818 m/s

In an elastic collision, kinetic energy is conserved as well as momentum, Consider a mass m moving with speed v_{i1} colliding with a mass M at rest. Letting the final speeds of m and M being v_{f1} and v_{f2}, one has

$$\frac{1}{2}mv_{i1}^2 = \frac{1}{2}mv_{f1}^2 + \frac{1}{2}Mv_{f2}^2,$$

and

$$mv_{i1} = mv_{i2} + Mv_{f2}.$$

Example 8.5 A mass m moving at speed v_{i1} strikes a mass M at rest in a head-on elastic collision. Calculate the speed of each of the masses after the collision.

In[7]:= **ClearAll["Global`*"];**

$Assumptions = $v_{f2} \neq 0$;

Solve $\left[m \, v_{i1} == m \, v_{f1} + M \, v_{f2} \,\,\&\&\,\, \frac{1}{2} m \, v_{i1}^2 == \frac{1}{2} m \, v_{f1}^2 + \frac{1}{2} M \, v_{f2}^2, \right.$

$\left. \{v_{f1}, v_{f2}\} \right]$

Out[7]= $\left\{ \left\{ v_{f1} \to \frac{(m - M) \, v_{i1}}{m + M}, \,\, v_{f2} \to \frac{2 \, m \, v_{i1}}{m + M} \right\} \right\}$

There are several interesting limits in Ex. 8.5. For $M \gg m$, one gets

$$v_{f1} = -v_{i1}$$

and

$$v_{f2} = 0.$$

For $m = M$, one gets

$$v_{f1} = 0$$

and

$$v_{f2} = v_{f1}.$$

For $m \gg M$, one gets

$$v_{f1} = v_{i1}$$

and

$$v_{f2} = 2v_{i1}.$$

Example 8.6 Two balls with masses m and M are dropped together and bounce off the floor in an elastic collision. Find the speed of the upper ball and the mass of the lower ball if the lower ball is at rest after the collision.

In[8]:= **\$Assumptions = {V > 0, v > 0, M > 0, m > 0};**

Solve$\left[M v - m v \; == m V \; \&\& \; \dfrac{1}{2} m v^2 + \dfrac{1}{2} M v^2 \; == \; \dfrac{1}{2} m V^2, \right.$

$\left. \{V, M\} \right]$

Out[8]= **{{V → 2 v, M → 3 m}}**

8.4.1 Ballistic Pendulum

In a ballistic pendulum, a projectile of mass m is fired into a hanging block of mass M where it suffers a completely inelastic collision. After the collision, the combined object swings up conserving energy. Conservation of momentum in the collision gives

$$mv = (m + M)V,$$

where v is the velocity of the projectile and V is the velocity after the collision. Conservation of energy after the collision gives

$$\frac{1}{2}(m + M)V^2 = (m + M)gh,$$

if the pendulum swings up a vertical distance h.

Example 8.7 In a ballistic pendulum, the projectile mass is 20 g and the pendulum mass is 1 kg. The combined mass swings up a vertical distance of 10 cm. Calculate the projectile speed.

In[9]:= $Assumptions = {V > 0, v > 0, M > 0, m > 0, g > 0, h > 0};

sol = Solve[m v == (m + M) V && $\frac{1}{2}$ (m + M) V² == (m + M) g h,

{v, V}] // Simplify

UnitConvert[v /. sol[[1]] /.

{h → 0.1 m, g → g, m → 20 g, M → 1 kg}

Out[9]= $\left\{\left\{v \to \frac{\sqrt{2}\ \sqrt{g\,h}\ (m + M)}{m}, V \to \sqrt{2}\ \sqrt{g\,h}\right\}\right\}$

Out[10]= 71.4242 m/s

8.4.2 Billiard Balls

Consider the elastic scattering of two billiard balls of equal mass m (Figure 8.2), After the collision, the balls emerge from the collision at angles θ and ϕ. Conservation of energy gives

$$\frac{1}{2}mv_{i1}^2 = \frac{1}{2}mv_{f1}^2 + \frac{1}{2}mv_{f2}^2.$$

Conservation of momentum gives

$$mv_{f1}\sin\theta = mv_{f2}\sin\phi,$$

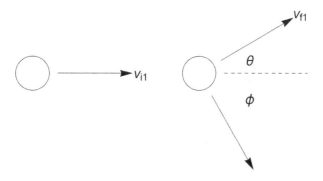

Figure 8.2   In the elastic scattering of two billiard balls of equal mass, the balls emerge from the collision at angles θ and ϕ.

and

$$mv_{i1} = mv_{f1}\cos\theta + mv_{f2}\sin\phi,$$

Example 8.8 Solve for the final velocities of the two billiard balls as a function of scattering angle.

In[11]:= $Assumptions = \left\{v_{i1} > 0, v_{f1} > 0, v_{f2} > 0, 0 < \theta < \frac{\pi}{2},\right.$

$$\left. 0 < \phi < \frac{\pi}{2}, m > 0\right\};$$

sol =

Quiet$\left[\right.$

Solve$\left[\frac{1}{2} m v_{i1}^2 = \frac{1}{2} m v_{f1}^2 + \frac{1}{2} m v_{f2}^2 \&\& \right.$

$m v_{f1} Sin[\theta] == m v_{f2} Sin[\phi] \&\&$

$m v_{i1} == m v_{f1} Cos[\theta] + m v_{f2} Cos[\phi], \{v_{f1}, v_{f2}, \phi\}\Big] //$

FullSimplify$\left.\right]$

Out[11]= $\left\{\left\{v_{f1} \to Cos[\theta] v_{i1}, v_{f2} \to Sin[\theta] v_{i1},\right.\right.$

$$\left.\left.\phi \to -2 ArcTan\left[1 - \frac{2}{1 + Tan\left[\frac{\theta}{2}\right]}\right]\right\}\right\}$$

Example 8.9 Evaluate $\sin\phi$.

In[12]:= $Sin\left[-2 ArcTan\left[1 - \frac{2}{1 + Tan\left[\frac{\theta}{2}\right]}\right]\right] //$ FullSimplify

Out[12]= $Cos[\theta]$

Since $\sin\phi = \cos\theta$, it must be true that $\theta + \phi = \pi/2$, or $90°$.

Example 8.10 Verify that $\sin\left(\frac{\pi}{2} - \theta\right) = \cos\theta$.

In[13]:= $Sin\left[\frac{\pi}{2} - \theta\right] == Cos[\theta]$

Out[13]= True

Rotational Motion

9.1 ANGLE

An angle θ (Figure 9.1) is defined by the arc length s divided by the radius, r

$$\theta = \frac{s}{r}.$$

Angles are dimensionless and this dimensionless quantity is often called a radian. When you see the word radian it does not mean anything in terms of a unit except to remind you that the angle is not in degrees. One complete trip (360°) around a circle of unit radius is an angle of 2π.

Example 9.1 Calculate the arc length of a 20-degree portion of a circle with a radius of 20 m.

```
In[1]:= UnitConvert[(20 Degree) (20. m)]

Out[1]= 6.98132 m
```

9.2 ANGULAR VELOCITY

We define angular velocity and acceleration just as was done for linear motion in Chap. 4. An angular displacement (Figure 9.2) is given by

$$\Delta\theta = \theta_2 - \theta_1.$$

If this displacement occurs over a time interval $t_2 - t_1$, then the average angular velocity in that interval has a magnitude

$$\bar{\omega} = \frac{\theta_2 - \theta_1}{t_2 - t_1} = \frac{\Delta\theta}{\Delta t}.$$

DOI: 10.1201/9781003481980-9

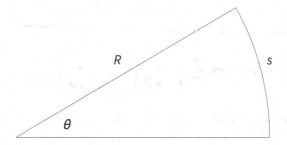

Figure 9.1 The measure of an angle θ is the arc length s divided by the radius R.

In the limit where $\Delta t \to 0$, the average angular velocity becomes the instantaneous angular velocity,

$$\omega = \frac{d\theta}{dt}.$$

The direction of the angular velocity as well as the angle itself are taken to be perpendicular to the plane of the circle and given by the right-hand-rule (putting the fingers of the right hand along the arc defined by θ, the thumb gives the direction of the angular velocity, choosing between the two directions of "up" or "down").

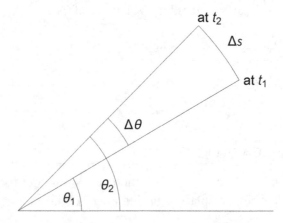

Figure 9.2 The average angular velocity is the change in angle divided by the time interval for the angle to change. The direction of the angular velocity is out of the page.

Example 9.2 Calculate the angular velocity of the earth's rotation.

In[2]:= $\textbf{UnitConvert}\left[\dfrac{2\,\pi}{24.\,\textbf{h}}\right]$

Out[2]= 0.0000727221 per second

9.3 ANGULAR ACCELERATION

If the angular velocity is changing we compute the average angular acceleration as

$$\bar{\alpha} = \frac{\omega_2 - \omega_1}{t_2 - t_1} = \frac{\Delta\omega}{\Delta t}.$$

In the limit where $\Delta t \to 0$, the average angular acceleration becomes the instantaneous angular acceleration

$$\alpha = \frac{d\omega}{dt}.$$

In summary, we have

$$s = \theta r,$$

$$v = \frac{dx}{dt} = r\frac{d\theta}{dt} = r\omega,$$

and

$$a = \frac{dv}{dt} = r\frac{d\omega}{dt} = r\alpha.$$

This means that

$$\omega = \int dt\,\alpha,$$

and

$$\theta = \int dt\,\omega.$$

9.4 CONSTANT ANGULAR ACCELERATION

If the acceleration is constant,

$$\omega = \omega_0 + \alpha t,$$

where ω_0 is the initial angular velocity, and

$$\theta = \theta_0 + \omega_0 t + \frac{1}{2}\alpha t^2,$$

where θ_0 is the initial angle.

Example 9.3 A phonograph record has an initial angular speed of 33 1/3 revolutions per minute. The record comes to a stop with constant angular acceleration in 6 s. Calculate the angular acceleration.

In[3]:= $\omega_0 = \dfrac{100}{3} \dfrac{2\pi}{\text{min}}$; t = 6. s;

sol = Solve[ω == ω_0 + α t, {α}] /. ω → 0 s^{-1};

α = UnitConvert[α /. sol[[1]]]

Out[5]= -0.581776 per $second^2$

Notice that in Ex. 9.3, one needs to put the angular velocity in rad/s which means s^{-1}, and one revolution corresponds to 2π radians.

Example 9.4 A bug rides on the spinning record of Ex. 9.3 at a distance of 0.3 m from the rotation axis as it stops. How far does the bug travel?

In[6]:= r = 0.3 m; $r\left(\omega_0 \, t - \dfrac{1}{2} \alpha \, t^2\right)$

Out[6]= 9.42478 m

9.5 MOMENT OF INERTIA

If we have a rigid rotating object, then every piece of the object rotates with the same angular velocity ω. Each piece of the object (m_i), however, is in general at a different distance. The total kinetic energy of the rotating object is the sum of the pieces,

$$K = \sum_i K_i = \sum_i \frac{1}{2} m_i v_i^2 = \omega^2 \sum_i \frac{1}{2} m_i r_i^2 .$$

The moment of inertia (I) is defined to be

$$I \equiv \sum_i \frac{1}{2} m_i r_i^2 ,$$

to give

$$K = \frac{1}{2} I \omega^2 .$$

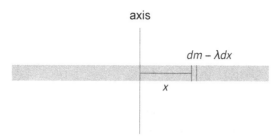

axis

$dm - \lambda dx$

x

Figure 9.3 The moment of inertia for a stick rotating about an axis through its midpoint is calculated by weighting each piece of the stick by its distance squared.

If we have a continuous distribution of matter, then the summation of the pieces becomes an integral, and one has

$$I = \int dm \, r^2.$$

The moment of inertia depends on the location of the rotation axis.

9.5.1 Stick

Consider a thin stick of material (Figure 9.3). To calculate the moment of inertia, one needs to relate the distribution of material to the distance from the rotation axis,

$$dm = \lambda dx,$$

where λ is the linear density (kg/m) of the rod. A uniform stick corresponds to constant λ.

Example 9.5 Calculate the moment of inertia for a uniform stick rotating about its center.

$\text{In[8]:=} \quad \int_{-\frac{L}{2}}^{\frac{L}{2}} \lambda \, x^2 \, dx \; / . \; \lambda \rightarrow \frac{M}{L}$

$\text{Out[8]=} \quad \dfrac{L^2 M}{12}$

Example 9.6 Calculate the moment of inertia for a uniform stick rotating about its end.

In[9]:= $\int_{0}^{L} \lambda \, x^2 \, dx \; / . \; \lambda \rightarrow \dfrac{M}{L}$

Out[9]= $\dfrac{L^2 M}{3}$

Note that the moment of inertia for rotation through the end-axis is larger than through the center-axis by an amount

$$\frac{ML^2}{3} - \frac{ML^2}{12} = M\left(\frac{L}{2}\right)^2.$$

It is generally true that the moment of inertia about an axis that is parallel ($I_{\parallel}$) to an axis passing through the center of mass (I_{cm}) is larger by an amount equal to MR^2, where R is the distance between the two parallel axis. This is known as the parallel axis theorem,

$$I_{\parallel} = I_{cm} + MR^2.$$

One can readily calculate the moment of inertia for nonuniform densities by inserting the density.

Example 9.7 Calculate the moment of inertia for a nonuniform stick rotating about its end if the density varies with distance from the end as $\lambda = (2 \text{ kg/m}^2)x$.

In[10]:= $L = 1 \text{ m}; \; \lambda = 2 \; \dfrac{\text{kg}}{\text{m}^2} \; \text{x}; \; \int_{0 \, m}^{L} \lambda \, x^2 \, dx$

Out[10]= $\dfrac{1}{2} \, \text{kg m}^2$

9.5.2 Cylinder

A cylindrical shell of mass M and radius R rotating about its axis has a simple moment of inertia because all the mass is at the same distance,

$$I = MR^2.$$

For a solid cylinder, one needs to integrate. For uniform volume density ρ,

$$dm = \rho dV = 2\pi r dr L\rho.$$

The answer does not depend on the length.

Example 9.8 Calculate the moment of inertia of a solid cylinder of mass M and radius R rotating about its axis.

In[11]:= **ClearAll["Global`*"];** $\int_0^R 2\pi\, r\, L\, \rho\ r^2\, dr\, /.\, \rho \to \dfrac{M}{\pi\, R^2\, L}$

Out[11]= $\dfrac{M\, R^2}{2}$

For a rotation axis along a diameter of the cylinder and passing through the center of mass, the distance to the rotation axis is

$$x = \sqrt{(r\sin\phi)^2 + z^2},$$

where ϕ is the azimuthal angle, r is the radial variable, and z runs along the axis of the cylinder, the usual cylindrical coordinates (App. B.2).

Example 9.9 Calculate the moment of inertia of a solid cylinder of mass M and radius R rotating about an axis along a diameter passing through the center of the cylinder.

In[2]:= $\int_0^{2\pi} \int_0^R \int_{-L/2}^{L/2} ((r\ \mathbf{Sin}[\phi])^2 + z^2)\ r\ \rho\ \mathbf{dz\, dr\, d\phi}\ /.\, \rho \to \dfrac{M}{\pi\, R^2\, L}$

Out[2]= $\dfrac{1}{12}\, M\, (L^2 + 3\, R^2)$

9.5.3 Disk

For a disk about along its axis perpendicular to the disk, the answer is the same as for a cylinder. For an axis along its diameter, the distance to the rotation axis is $r\sin\phi$, where ϕ is the angle measured w.r.t. the axis.

Example 9.10 Calculate the moment of inertia of a disk of mass M and radius R rotating about an axis along its diameter.

In[40]:= $\int_0^R \int_0^{2\pi} (r\ \mathbf{Sin}[\phi])^2\ r\ \sigma\, \mathbf{d\phi\, dr}\ /.\, \sigma \to \dfrac{M}{\pi\, R^2}$

Out[40]= $\dfrac{M\, R^2}{4}$

Note that for $L = 0$, the answer from Ex. 9.9 is reproduced.

axis

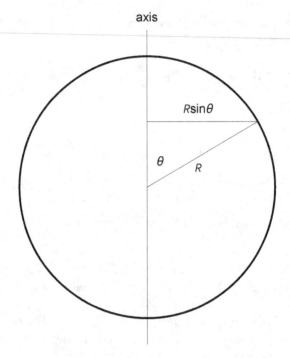

Figure 9.4 The moment of inertia for a sphere rotating about an axis through its midpoint is calculated by weighting the each piece of the sphere by its distance squared to the rotation axis.

9.5.4 Sphere

For a sphere (shell) of mass M and radius R, rotating about an axis though its center (Figure 9.4) the distance to the rotation axis for an arbitrary point of the sphere point on the sphere is $R\sin\theta$ where θ is the polar angle (App. B.1). Thus, the moment of inertia is given by

$$I = 2\pi \int_0^\pi d\theta \, (R\sin\theta)^2 \sigma R^2 \sin\theta,$$

where $\sigma = M/(4\pi R^2)$, the mass per area on the sphere.

Example 9.11 Calculate the moment of inertia of a (hollow) sphere of mass M and radius R.

In[12]:= $2\pi \int_{0}^{\pi} (R \; Sin[\theta])^2 \; R^2 \; Sin[\theta] \; \sigma \, d\theta \; /. \; \sigma \rightarrow \dfrac{M}{4\pi \; R^2}$

Out[12]= $\dfrac{2\,M\,R^2}{3}$

The moment of inertia of a sphere about an axis that is tangent to the sphere may be calculated with the aid of Figure 9.5. The distance to the rotation axis (x) now depends on both the polar angle θ and the azimuthal angle ϕ. The segment $R\sin\theta$ reaches from the center to an arbitrary position on the sphere and together with the segment R that reaches to the rotation axis and x make a triangle. Note that this triangle is not in the plane that is perpendicular to the rotation axis. Using the law of cosines (Section 3.4), the distance to the rotation axis is

$$x = \sqrt{R^2 + (R\sin\theta)^2 - 2R^2 \sin\theta\cos\phi}.$$

Example 9.12 Calculate the moment of inertia of a sphere of mass M and radius R about an axis that is tangent to the sphere.

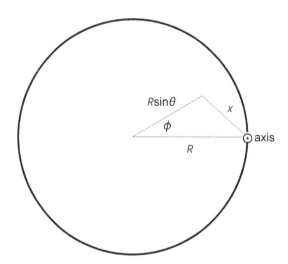

Figure 9.5   The distance triangle is shown for the rotation axis tangent to the sphere. The triangle is not in the plane perpendicular to the rotation axis. The drawing is in the ϕ plane and the segment $R\sin\theta$ is not in that plane unless $\theta = \pi/2$.

In[13]:= $\int_{0}^{2\pi} \int_{0}^{\pi} (R^2 + (R \, Sin[\theta])^2 - 2\,R\,(R\,Sin[\theta])\,Cos[\phi]) \, R^2$

$$Sin[\theta] \, \sigma \, d\theta \, d\phi \, /. \, \sigma \rightarrow \frac{M}{4\,\pi\,R^2}$$

Out[13]= $\dfrac{5\,M\,R^2}{3}$

The parallel axis theorem is seen to hold,

$$\frac{5}{3}MR^2 = \frac{2}{3}MR^2 + MR^2.$$

9.5.5 Ball

For a sphere (shell) of mass M and radius R, rotating about an axis though its center, the distance to the rotation axis for a point inside the ball is $r\sin\theta$ (Figure 9.6). The moment of inertia is given by

$$I = 2\pi \int_0^\pi d\theta \, (r\sin\theta)^2 \rho r^2 \sin\theta,$$

where $\rho = M/(\frac{4}{3}\pi R^3)$, the mass per volume of the ball.

Example 9.13 Calculate the moment of inertia of a uniform ball of mass M and radius R.

In[14]:= $2\pi \int_{0}^{R} \int_{0}^{\pi} (r \, Sin[\theta])^2 \, r^2 \, Sin[\theta] \, \rho \, d\theta \, dr \, /. \, \rho \rightarrow \dfrac{M}{\frac{4}{3}\pi\,R^3}$

Out[14]= $\dfrac{2\,M\,R^2}{5}$

9.6   CONSERVATION OF ENERGY

Consider a string wrapped around a cylinder of radius R with a string wrapped around its outside edge (Figure 9.7). If the cylinder has an angular speed ω, it will climb up the string. The height h that it will reach is given by energy conservation,

$$\frac{1}{2}I\omega^2 = Mgh.$$

The mass will cancel out because it is also contained within the moment of inertia I.

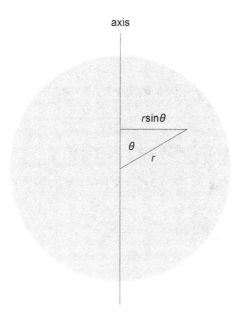

Figure 9.6 The moment of inertia for a ball rotating about an axis through its midpoint is calculated by weighting the each piece of the ball by its distance squared to the rotation axis.

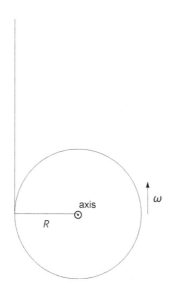

Figure 9.7 A cylinder with an angular speed ω climbs a string wrapped around its outer edge.

Example 9.14 A cylinder of mass M and radius R has an angular speed ω. What angular speed is needed to climb to a height of h? Give a numerical answer for $R = 3$ cm and $h = 1$ m.

```
In[34]:= $Assumptions = {ω > 0, R > 0, h > 0, g > 0};

        I = 1/2 M R²;

        sol = Solve[1/2 I ω² == M g h, {ω}]

        ω /. sol[[1]] /. {R → 3. cm, h → 1 m, g → g};

        UnitConvert[%]
```

$$\text{Out[34]= } \left\{\left\{\omega \to \frac{2\sqrt{g\,h}}{R}\right\}\right\}$$

Out[36]= 208.77 per second

Consider a wheel that has a moment of inertia I and radius of R from which a mass is hanging from a string wrapped around the perimeter of the wheel (Figure 9.8). As the mass falls, the wheel turns. Energy is conserved. The key relationship between the angular speed of the wheel and the linear speed of the mass m, due to the attachment of the string, is

$$\omega = \frac{v}{R}.$$

Example 9.15 Calculate the speed of the mass m after it has dropped a height h. Calculate the acceleration of the falling mass.

```
In[24]:= $Assumptions = {ω > 0, v > 0, R > 0, g > 0, h > 0,
         M > 0, I > 0};

        sol = Solve[1/2 M v² + 1/2 I ω² == M g h && ω == v/R, {v, ω}]

        a = (v /. sol[[1]])²/(2 h)
```

$$\text{Out[24]= } \left\{\left\{v \to \frac{1}{5}\sqrt{34}\,R\sqrt{\frac{g\,M}{M\,R^2 + I}}, \; \omega \to \frac{1}{5}\sqrt{34}\sqrt{\frac{g\,M}{M\,R^2 + I}}\right\}\right\}$$

$$\text{Out[25]= } \frac{g\,M\,R^2}{M\,R^2 + I}$$

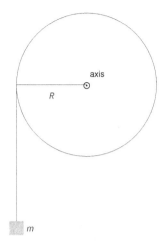

Figure 9.8   A wheel of radius R has a string wrapped around its perimeter with a mass m attached. As the mass falls, the wheel turns and energy is conserved.

9.7   TORQUE

9.7.1   Definition

The ability to make an object rotate depends on both the applied force and its position and direction relative to the rotation axis (Figure 9.9). The definition of torque (τ) is

$$\boldsymbol{\tau} = \mathbf{r} \times \mathbf{F} = rF \sin\phi,$$

where $\mathbf{r}$ is the position vector that points from the rotation axis to the point where the force ($\mathbf{F}$) is applied and ϕ is the angle between the vectors $\mathbf{r}$ and $\mathbf{F}$.

Writing out components

$$\mathbf{r} = r_x\,\hat{\mathbf{i}} + r_y\,\hat{\mathbf{j}} + r_z\,\hat{\mathbf{k}}$$

and

$$\mathbf{F} = F_x\,\hat{\mathbf{i}} + F_y\,\hat{\mathbf{j}} + F_z\,\hat{\mathbf{k}}$$

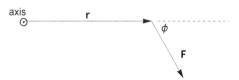

Figure 9.9   A force $\mathbf{F}$ acts at a position $\mathbf{r}$.

one gets

$$\tau = (r_y F_z - r_z F_y)\,\hat{\mathbf{i}} + (r_z F_x - r_x F_z)\,\hat{\mathbf{j}} + (r_x F_y - r_y F_z)\,\hat{\mathbf{k}}$$

Example 9.16 A force $\mathbf{F} = 4\,\text{N}\,\hat{\mathbf{i}} + 5\,\text{N}\,\hat{\mathbf{j}} + 6\,\text{N}\,\hat{\mathbf{k}}$ acts at a position $\mathbf{r} = 1\,\text{m}\,\hat{\mathbf{i}} + 2\,\text{m}\,\hat{\mathbf{j}} + 3\,\text{m}\,\hat{\mathbf{k}}$. Calculate the torque vector.

In[50]:= **F = {4, 5, 6} N; r = {1, 2, 3} m; r × F**

Out[50]= $\left\{-3\,\text{mN}, \ 6\,\text{mN}, \ -3\,\text{mN}\right\}$

9.7.2 Newton's Second Law

Consider a mass m undergoing nonuniform circular motion with radius R due to a tangential component of the force ($\mathbf{F}_t$), The tangential force causes a tangential acceleration a_t,

$$F_t = ma_t.$$

The magnitude of the torque is

$$\tau = RF_t = Rma_t = mR^2\alpha = I\alpha.$$

The torque and acceleration vectors are in the same direction, and Newton's 2nd law in terms of torque reads

$$\tau = I\alpha.$$

9.8 ANGULAR MOMENTUM

Example 9.15 can be worked with conservation of angular momentum. For the falling mass,

$$mg - T = ma.$$

For the rotating wheel, using $\tau = I\alpha$,

$$TR = I\alpha = \frac{Ia}{R},$$

resulting in the same acceleration as calculated above.

9.9 ROLLING WITHOUT SLIPPING

Rolling without slipping may be viewed as a superposition of translational and rotational motion (Figure 9.10). The rolling object is both moving as a whole (with no rotation) and rotating. The speeds are related by

$$v_{\text{rot}} = v_{\text{cm}}.$$

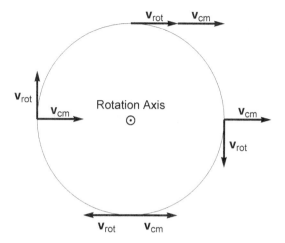

Figure 9.10 A wheel rotates about its center of mass as it rolls on a level surface.

The direction of v_{cm} is constant while the direction of v_{rot} is tangent to the rim of the wheel. The top of the wheel is moving two times faster than the translational speed. The bottom of the wheel at the point which it is in contact with the surface, is stationary. The force that causes the torque to rotate the wheel is static friction. Figure 9.11 shows the magnitude of the speed of a point on the rim of the wheel vs. θ.

Consider an object rolling without slipping down an inclined plane (Figure 9.12). Consider objects of various shapes that have a radius R. The speed and acceleration can be solved by energy conservation. After dropping a ver-

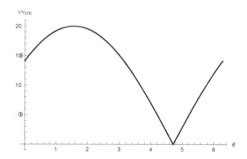

Figure 9.11 The magnitude of the speed of a point on the rim of a wheel rotating on a level surface without slipping, relative to the translational speed. The angle starts at zero at the "three o'clock" position and runs counter clockwise.

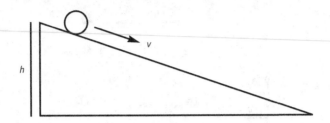

Figure 9.12 An object with a moment of inertia I, for rotation about its center of mass, rolls without slipping down an inclined plane.

tical distance h,

$$mgh = \frac{1}{2}mv^2 + \frac{1}{2}I\left(\frac{v}{R}\right)^2,$$

or

$$v = \sqrt{\frac{2mgh}{m + I/R^2}}.$$

The acceleration is

$$a = \frac{v^2}{2h}.$$

Example 9.17 Calculate the expression for the speed for rolling without slipping after dropping a vertical height h for a hoop, sphere, solid cylinder, and ball. Give numerical answers for $h = 1$ m.

```
Clear[v, I];

v[I_] := √( (2 m g h) / (m + I/R²) );

{v[m R²], v[2/3 m R²], v[1/2 m R²], v[2/5 m R²]}
UnitConvert[% /. {h → 1. m, g → g}]
```

Out[48]= $\left\{ \sqrt{g\,h},\ \sqrt{\frac{6}{5}}\ \sqrt{g\,h},\ \frac{2\sqrt{g\,h}}{\sqrt{3}},\ \sqrt{\frac{10}{7}}\ \sqrt{g\,h} \right\}$

Out[49]= $\{ 3.13156\,\text{m/s},\ 3.43045\,\text{m/s},\ 3.61601\,\text{m/s},\ 3.74293\,\text{m/s} \}$

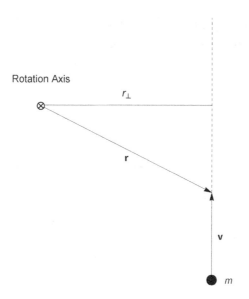

Figure 9.13 The angular momentum with respect to a predefined rotation axis of a mass m moving with speed v is $mvr_\perp$, where $r_\perp$ is the perpendicular distance to the rotation axis.

9.10 CONSERVATION OF ANGULAR MOMENTUM

For motion in a straight line, the angular momentum about a predefined axis is the momentum times the perpendicular distance to that axis (Figure 9.13),

$$L = pr_\perp = mvr_\perp.$$

This may be written in terms of a vector cross product as

$$\mathbf{L} = \mathbf{r} \times \mathbf{p}.$$

Consider a playground "merry-go-round" or carousel with moment of inertia I, radius R, and angular speed ω_i. A child with mass m runs with speed v along a trajectory that intersects the rotation axis and jumps to the edge of the rotating carousel. Conservation of angular momentum gives

$$I\omega_i + 0 = (I + mR^2)\omega_f,$$

where ω_f is the speed of the rotating system (carousel plus child). This gives

$$\omega_f = \frac{I\omega_i}{I + mR^2}.$$

The resulting angular speed is smaller.

Now, suppose the child has a different trajectory that is tangent to the edge of the carousel. Conservation of angular momentum now reads

$$I\omega_i + Rmv = (I + mR^2)\omega_f,$$

and

$$\omega_f = \frac{I\omega_i + Rmv}{I + mR^2} = \frac{I + \frac{mvR}{\omega_i}}{I + mR^2}.$$

One can see that if $v > \omega_i R$, then the system speeds up after the child jumps on. If $v < \omega_i R$, then the rotating system slows down.

Example 9.18 A carousel has $I = 400$ kg · m², $R = 2$ m, and $\omega_i = 2$ s^{-1}. A child with mass 50 kg jumps on tangent to the edge with a speed of 5 m/s. Calculate the angular speed of the rotating system.

```
In[24]:= I = 400. kg m²; R = 2 m; ωᵢ = 2 s⁻¹;

                                                   I + m v R
                                                       ωᵢ
        v = 5 m/s; m = 50 kg; UnitConvert[ ───────────── ] ωᵢ
                                              I + m R²


Out[25]= 2.16667 per second
```

CHAPTER 10

Universal Gravitation

10.1 UNIVERSAL FORCE LAW

The universal force of attraction between two masses m_1 and m_2 is

$$F = G\frac{m_1 m_2}{r^2},$$

where r is their separation distance (center-to-center). For a mass m attracted to the earth (M_e) near its surface,

$$mg = G\frac{m M_E}{R_E^2},$$

where R_e is the radius of the earth. Thus,

$$g = G\frac{M_E}{R_E^2}.$$

Example 10.1 Get the universal gravitational constant (G).

In[1]:= **Quantity["GravitationalConstant"]**

Out[1]= G

In[2]:= **UnitConvert$\left[G\right]$**

Out[2]= $6.674 \times 10^{-11}\,\mathrm{m^3/(kg\,s^2)}$

DOI: 10.1201/9781003481980-10

Example 10.2 Get the mass of the earth.

In[3]:= **Quantity["EarthMass"]**

Out[3]= $M_\oplus$

In[4]:= **UnitConvert$\left[M_\oplus\right]$**

Out[4]= 5.97×10^{24} kg

Example 10.3 Get the radius of the earth.

In[7]:= **R_e = Quantity["EarthEquatorialRadius"]**

Out[7]= a

In[8]:= **UnitConvert[R_e]**

Out[8]= 6 378 137 m

Example 10.4 Calculate the acceleration of gravity (g) at the surface of the earth.

In[10]:= **UnitConvert$\left[\dfrac{G\ M_\oplus}{R_e{}^2}\right]$**

Out[10]= 9.80 m/s^2

10.2 POTENTIAL ENERGY

The change in gravitational potential energy is written

$$\Delta U = -\int_{r_a}^{r_b} dr\, F = -GmM \int_{r_a}^{r_b} \frac{dr}{r^2} = GmM\left(\frac{1}{r_b} - \frac{1}{r_a}\right).$$

The zero point may be chose at infinity to give

$$U(r) = -\frac{GmM}{r}.$$

The gravitational potential energy for a mass m at the surface of the earth is

$$U = -G\frac{mM_e}{R_e}.$$

Example 10.5 Show that for heights h above the earth's surface that are small compared to its radius, the gravitational potential energy reduces to mgh.

In[8]:= **Series**$\left[-\dfrac{G\, m\, M}{R+h} - \left(-\dfrac{G\, m\, M}{R}\right), \{h,\, 0,\, 1\}\right]$ /. $\dfrac{G\, M}{R^2} \to g$

Out[8]= $g\, m\, h + O[h]^2$

Example 10.6 Calculate the escape velocity, the speed at which an object needs to be launched in order to escape the gravitational pull of the earth.

In[9]:= $K_i = \dfrac{1}{2}\, m\, v_i{}^2;\; U_i = -\dfrac{G\, m\, M_e}{R_e};\; K_f = 0;\; U_f = 0;$

Solve$[K_i + U_i == K_f + U_f,\, v_i]$ // **Quiet**

Out[10]= $\left\{\left\{v_i \to -1.118\times10^4 \text{ m/s}\right\},\, \left\{v_i \to 1.118\times10^4 \text{ m/s}\right\}\right\}$

Only the positive solution is physical.

10.3 GRAVITATIONAL FORCE FROM A SPHERE

This section shows how to calculate gravitational force from extended objects. This section concentrates on a sphere, but Mathematica is capable of calculating the force from any shape. As a stepping stone to understanding how to calculate the gravitational force from a sphere, one may start with the simpler case of a ring.

10.3.1 Ring of Mass

Consider a ring of mass described by some linear density λ in units of kg/m. The total mass of the ring is

$$M = 2\pi R\lambda.$$

On the axis of the ring, at a distance r from the center, the gravitational force between a point mass m and the ring is simple because every point on the ring is at the same distance. From symmetry, one knows that the force is along the axis of the ring. Therefore, one must put in the factor that gives that component,

$$\cos\theta = \frac{r}{\sqrt{r^2 + R^2}}.$$

The gravitational force on m caused by the ring of mass is

$$\mathbf{F} = \frac{GmM}{(r^2 + R^2)} \cos\theta(-\hat{\mathbf{r}}) = \frac{GmMr}{(r^2 + R^2)^{3/2}}(-\hat{\mathbf{r}}),$$

where $\hat{\mathbf{r}}$ points in the direction from the center of the ring to the mass m.

It is useful to set up a coordinate system where the origin is at the center of the ring (Figure 10.1). Define the vector $\mathbf{r}$ to point from the origin to the place where the force is being calculated, in this case the location of mass m. Define a second vector $\mathbf{r}'$, which will become a summation or integration variable that points to the geometric object. Define a third vector $\mathcal{R}$ that points from a tiny piece of the geometric shape to the location of m.

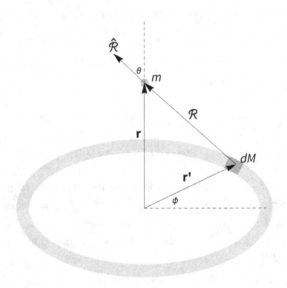

Figure 10.1　A mass m is attracted to a ring of mass M. The contribution to the force from dM is along $(-\hat{\mathcal{R}})$.

The three vectors make a triangle,

$$\mathbf{r}' + \boldsymbol{\mathcal{R}} = \mathbf{r},$$

or

$$\boldsymbol{\mathcal{R}} = \mathbf{r} - \mathbf{r}'.$$

The contribution to the force ($d\mathbf{F}$) on m from a tiny piece of the ring (dM) is

$$d\mathbf{F} = \frac{GmdM}{\mathcal{R}^2}(-\hat{\boldsymbol{\mathcal{R}}}) = -\frac{GmdM}{\mathcal{R}^3}\boldsymbol{\mathcal{R}}. = -\frac{GmdM}{(\boldsymbol{\mathcal{R}}\cdot\boldsymbol{\mathcal{R}})^{3/2}}\boldsymbol{\mathcal{R}}$$

The final step in the set up is to relate dm to the specified mass density,

$$dM = \lambda R d\phi,$$

where ϕ is the angular variable (our integration variable) that sweeps out the ring of mass. Thus,

$$\mathbf{F} = \int d\mathbf{F} = -GmR\lambda \int_0^{2\pi} \frac{d\phi\,\boldsymbol{\mathcal{R}}}{(\boldsymbol{\mathcal{R}}\cdot\boldsymbol{\mathcal{R}})^{3/2}}.$$

Example 10.7 Calculate the force on m from the ring of mass M.

In[11]:= $\mathcal{R}$ = {0, 0, r} - {R Cos[ϕ], R Sin[ϕ], 0};
$$\int_0^{2\pi} -\frac{G\,m\,R\,\lambda\,\mathcal{R}}{(\mathcal{R}.\mathcal{R})^{3/2}}\,d\phi\ /. \lambda \to \frac{M}{2\pi R}$$

Out[12]= $\left\{0, 0, -\dfrac{G\,m\,M\,r}{\left(r^2 + R^2\right)^{3/2}}\right\}$

10.3.2 Sphere of Mass

The attraction of a point mass to a sphere of mass may be calculated by dividing the sphere into rings (Figure 10.2). Consider a ring at an arbitrary angle θ. The radius of the sphere is $R\sin\theta$, the distance to the sphere is $R - R\cos\theta$. The mass density is $M/(4\pi R^2)$. The mass of a ring is

$$dM = (2\pi R)(R\sin\theta).$$

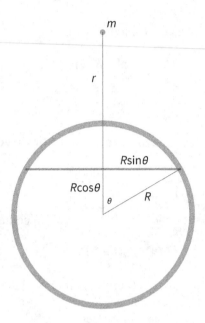

Figure 10.2 A mass m is attracted to a sphere of mass M. The problem can be solved by dividing the sphere into rings.

Example 10.8 Calculate the force on m from the sphere of mass M.

```
In[13]:= $Assumptions = {r > 0, R > 0};
```

$$I[\theta\_] = -G\, m \, \frac{M}{4\, \pi\, R^2} \, 2\, \pi\, R^2$$

$$\int \frac{\mathrm{Sin}[\theta]\ (r - R\, \mathrm{Cos}[\theta])}{\left((R\, \mathrm{Sin}[\theta])^2 + (r - R\, \mathrm{Cos}[\theta])^2\right)^{3/2}} \, d\theta;$$

```
I[π] - I[0] // Simplify
```

$$\mathrm{Out[14]=}\ \begin{cases} -\dfrac{G\, m\, M}{r^2} & r \geq R\ \&\&\ r^2 \geq R^2 \\ 0 & \mathrm{True} \end{cases}$$

This is a remarkable result. The gravitational force caused by the sphere behaves the same as if all the mass were concentrated at its center. Furthermore, if the mass m is inside the sphere, then the gravitational force is zero.

As an alternate solution, one can divide the surface of the sphere into little pieces of size $(R d\phi) \times (R\sin\theta d\theta)$ and then integrate over the surface of the sphere.

Example 10.9 Calculate the force on *m* from the sphere of mass *M* with direct integration.

```
In[15]:= ClearAll["Global`*"];
        $Assumptions = {r > 0, R > 0, θ' ∈ ℝ, φ' ∈ ℝ};
        r' = FromSphericalCoordinates[{R, θ', φ'}];
        ℛ = {0, 0, r} - r';
```

$$I_\phi[\phi'\_] = -\int \frac{(G\,m\,M)\,\mathcal{R}}{(4\,\pi)\,\sqrt{\mathcal{R}.\mathcal{R}}^3}\,d\phi';$$

$$I_\theta[\theta'\_] = \int (I_\phi[2\,\pi] - I_\phi[0])\,\text{Sin}[\theta']\,d\theta';$$

```
Simplify[Simplify[I_θ[π] - I_θ[0]]]
```

$$Out[18]= \left\{0,\ 0,\ \begin{cases} -\frac{G\,m\,M}{r^2} & r \geq R\ \&\&\ r^2 \geq R^2 \\ 0 & \text{True} \end{cases}\right\}$$

One could build up a ball of mass by adding up concentric spheres to see that the ball also behaves as if all the mass were concentrated at the center as concerns the attraction of an external mass.

10.3.3 Oscillating Mass

Suppose, a hole could be dug through the entire diameter of the earth (Figure 10.3).

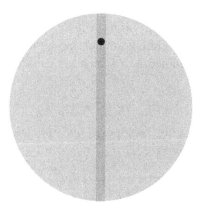

Figure 10.3 A mass is dropped into a hole that is dug through the diameter of the earth.

A mass dropped down that hole would only be pulled only be that portion of the mass whose radial distance is smaller than the mass. For a distance R, that mass is

$$M = \frac{4}{3}\pi r^3 \rho,$$

where ρ is the density of the earth, taken to be constant. The attractive force is

$$F = -\frac{GMm}{r^2} = -\frac{4}{3}\pi Gm\rho r = -kr,$$

where

$$k = -\frac{4}{3}\pi Gm\rho.$$

This is precisely the condition for simple harmonic motion.

Example 10.10 Calculate the period for the oscillating mass.

In[19]:= $k = \frac{4}{3} \pi\, G\, m\, \boxed{\text{Earth } \text{PLANET}}\, \left[\, \boxed{\text{mean density}}\, \right];$

$\text{UnitConvert}\left[2\, \pi\, \sqrt{\dfrac{m}{k}}\, ,\, \text{min}\right]$

Out[19]= $84.34\,\text{min}$

10.4   KEPLER'S LAWS

10.4.1   Law of Orbits

All planets move in elliptical orbits (Figure 10.4) with the sun at one focus. Consider a mass m orbiting a much more massive M with the distance between the masses denoted as r which varies with time. Conservation of energy gives

$$E = \frac{1}{2}mv^2 - \frac{GMm}{r}.$$

The velocity vector may be decomposed into radial (along the line connecting M and M) and transverse pieces,

$$v^2 = \left(\frac{dr}{dt}\right)^2 + r^2\omega^2 = \left(\frac{dr}{dt}\right)^2 + r^2\left(\frac{d\theta}{dt}\right)^2.$$

This allows the energy equation to be written

$$E = \frac{1}{2}m\left(\frac{dr}{dt}\right)^2 + \frac{1}{2}mr^2\left(\frac{d\theta}{dt}\right)^2 - \frac{GMm}{r}.$$

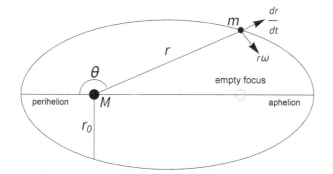

Figure 10.4 The orbit of mass m about mass M is the shape of an ellipse.

Conservation of angular momentum gives

$$L = mr^2\left(\frac{d\theta}{dt}\right).$$

It is convenient to make a change of variables,

$$\rho = \frac{1}{r},$$

to get

$$\theta = \frac{L}{m}\int \rho^2\, dt = \frac{L}{m}\int \rho^2\frac{dt}{d\rho}d\rho.$$

Note that

$$\frac{dr}{dt} = \frac{1}{\rho^2}\frac{d\rho}{dt},$$

which gives

$$\theta = -\frac{L}{m}\int \frac{d\rho}{dr/dt}.$$

Now solve the energy equation for dr/dt, using the expression for L to eliminate $d\theta/dt$,

$$\frac{dr}{dt} = \sqrt{\frac{2E}{m} + 2GM\rho - \frac{L^2\rho^2}{m^2}}.$$

Thus,

$$\theta = -\frac{L}{m}\int \frac{d\rho}{\sqrt{\frac{2E}{m} + 2GM\rho - \frac{L^2\rho^2}{m^2}}}.$$

To simplify the expression, define

$$r_0 = \frac{L^2}{GMm^2},$$

and

$$e = \sqrt{1 + \frac{2Er_0}{GMm}}.$$

The parameter r_0 is the radius for a circular orbit and it gives the distance at $\theta = \pi/2$ for an elliptical orbit. The parameter e (the eccentricity) specifies how r varies for an elliptical orbit.

Example 10.11 Solve for the relationship between θ and $\rho = 1/r$.

In[20]:= $Assumptions = {r₀ > 0, e > 0, ρ > 0, m > 0, M > 0,
    G > 0, L > 0, E > 0};

$$\text{rdot} = \sqrt{\frac{2\,E}{m} + 2\,G\,M\,\rho - \left(\frac{L\,\rho}{m}\right)^2}\,;$$

$$I = -\frac{L}{m\ \text{rdot}}\ /.\ \left\{L \to \sqrt{r_0\,G\,M\,m^2}\,,\ E \to \frac{(e^2 - 1)\,G\,M\,m}{2\,r_0}\right\}\ //$$

FullSimplify;

θ = Integrate[I, ρ] // Simplify

$$\text{Out[23]= } 2\,\text{ArcSin}\left[\frac{\sqrt{\dfrac{1 + e - \rho\,r_0}{e}}}{\sqrt{2}}\right]$$

The solution from Ex. 10.11 is that of an ellipse. To see that it is so, one can solve for $r = 1/\rho$ and then make a polar plot of (r, θ). Figure 10.5 shows the polar plot for $e = 0.7$.

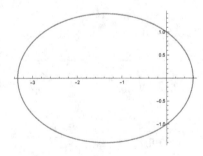

Figure 10.5 The polar plot of (r, θ) shows the path taken by the orbiting mass. The axis is at the focus and the angle θ is measured from the horizontal.

Example 10.12 Find the value of ρ for $\theta = \pi/2$. Solve for $r(\theta)$.

```
In[24]:= Solve[θ == π / 2, ρ]
        ClearAll["Global`*"];
        sol =
```

$$\text{NSolve}\left[\theta == 2 \ \text{ArcSin}\left[\frac{\sqrt{\frac{1+e-\rho\,r_\theta}{e}}}{\sqrt{2}}\right] \ /. \ \{e \rightarrow .7, \ r_\theta \rightarrow 1\},\right.$$

```
        ρ] // Quiet;
        r = (ρ /. sol[[1]])⁻¹
```

$$\text{Out[24]}= \left\{\left\{\rho \rightarrow \frac{1}{r_\theta}\right\}\right\}$$

$$\text{Out[27]}= \frac{1}{1.7 - 1.4 \, \text{Sin}[0.5\,\theta]^2}$$

10.4.2 Law of Areas

Planets sweep out equal areas A in equal times. For a short time Δt, the area swept out is a triangle (Figure 10.6), such that

$$\frac{dA}{dt} = \frac{\frac{1}{2}(r\Delta\theta)(r)}{\Delta t}.$$

In the limit that $\Delta t \rightarrow 0$,

$$\frac{dA}{dt} = \frac{1}{2}r^2\omega = \frac{L}{2m}.$$

Conservation of angular momentum guarantees the law of areas to hold true.

One can integrate to get the relationship between the total area of an orbit and the period.

$$A = \int dA = \frac{L}{2m} \int dt = \frac{LT}{2m}.$$

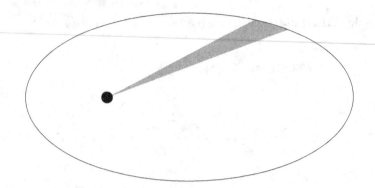

Figure 10.6 The area swept out by an elliptical orbit in a small time Δt is a triangle with height r and base $r\Delta\theta$.

Example 10.13 Calculate the angular momentum of the earth about the sun and use it to determine the period of the orbit.

In[28]:= M_e = Earth [*mass*];

v = Earth [*average orbit velocity*];

r = Earth [*orbital semimajor axis*]; A = πr^2;

L = M_e v r; T = UnitConvert$\left[\dfrac{2\,M_e\,A}{L}, \text{yr}\right]$

Out[29]= 1.00 yr

10.4.3 Law of Periods

The square of the orbit period is proportional to the average distance of the orbiting mass. For a circular orbit, Newton's 2nd law reads

$$\frac{GMm}{r^2} = \frac{mv^2}{r} = m\omega^2 r.$$

Using $\omega = 2\pi/T$,

$$T^2 = \left(\frac{4\pi^2}{GM}\right) r^3.$$

For an elliptical orbit, one may replace the radius by the semi-major axis a, which is related to the aphelion (R_a) and perihelion (R_p) distances by

$$R_a = a(1+e),$$

and
$$R_p = a(1-e).$$

The law of areas allows us to determine the mass of the central body (M) if one knows the radius and period of an orbiting mass m.

Example 10.14 Use the radius and period of the earth's orbit to determine the mass of the sun.

In[27]:= M = UnitConvert$\left[\dfrac{4\,\pi^2\,\text{au}^3}{G\,\text{yr}^2}\right]$

Out[27]= 1.9912×10^{30} kg

The law of areas relates the orbits of two objects (A and B) by
$$\frac{T_A^2}{r_A^3} = \frac{T_B^2}{r_B^3}.$$

The astronomical unit (au) is the mean distance between the earth and the sun. It can be used as a convenient yardstick to calculate the parameters of other orbits.

Example 10.15 Calculate the period of Pluto from its orbital semi-major axis.

In[28]:= $\sqrt{\dfrac{\boxed{\text{Pluto } \text{\tiny MINOR PLANET}}\,\boxed{\text{\textit{orbital semimajor axis}}}^3\,\text{yr}^2}{\text{au}^3}}$

Out[28]= 248.0810252 yr

Example 10.16 A satellite orbits the earth at a height of 250 km above the surface. Calculate the period.

In[29]:= M$_e$ = $\boxed{\text{Earth } \text{\tiny PLANET}}\,\boxed{\text{\textit{mass}}}$; h = 250 km;

r$_e$ = $\boxed{\text{Earth } \text{\tiny PLANET}}\,\boxed{\text{\textit{average radius}}}$ + h;

T = UnitConvert$\left[\sqrt{\dfrac{4\,\pi^2\,r_e^3}{G\,M_e}}\,,\ \text{min}\right]$

Out[30]= 89.4 min

A geosynchronous orbit is one in which its angular velocity matches the earth's rotation, in other words, its period is 24 h.

Example 10.17 Calculate the height (above the earth's surface) of a geosynchronous orbit.

$$\text{In[31]:= } T = 24 \text{ h}; r = \text{UnitConvert}\left[\left(\frac{G \; T^2 \; M_e}{4 \; \pi^2}\right)^{1/3}\right] - r_e$$

$$\text{Out[31]= } 3.562 \times 10^7 \text{ m}$$

Example 10.18 Halley's comet has a perihelion distance of 8.9×10^{10} m and a period of 76 years. Calculate the aphelion distance and the eccentricity of the orbit.

$$\text{In[32]:= } R_p = 8.9 \; 10^{10} \text{ m};$$

$$T = 76 \text{ yr}; M = \boxed{\text{Sun STAR}}\left[\boxed{mass}\right];$$

$$a = \text{UnitConvert}\left[\left(\frac{G \; T^2 \; M}{4 \; \pi^2}\right)^{1/3}\right];$$

$$R_a = 2 \; a \; - \; R_p$$

$$e = \frac{R_a \; - \; R_p}{2 \; a}$$

$$\text{Out[35]= } 5.2767 \times 10^{12} \text{ m}$$

$$\text{Out[36]= } 0.966826$$

The orbit of Halley's comet is a highly eccentric ellipse (Figure 10.7).

Figure 10.7 The orbit of Halley's comet is a highly eccentric ellipse.

CHAPTER 11

Fluids

A fluid is a substance that can flow. It can be a liquid or a gas.

11.1 DENSITY AND PRESSURE

Mass density (ρ) is mass per volume.

11.1.1 Water

Example 11.1 Get the density of water.

In[1]:= ρ = [**water** CHEMICAL] [*mass density*]

Out[1]= 0.9970480 g/cm^3

Pressure is the force per area. At a specified depth in a fluid, the pressure may be thought of as due to the weight of the fluid above it. Consider water at a depth h. The pressure calculated from the weight on an area A is

$$P = \frac{F}{A} = \frac{\rho V g}{A} = \rho g h.$$

Pressure is measured in the unit Pascal (Pa) which is 1 N/m^2.

Example 11.2 Get the pressure in water at a depth of 1 m.

In[2]:= **UnitConvert**$\left[\rho \ \mathbf{1} \ \mathbf{m} \ g, \ \mathbf{Pa}\right]$

Out[2]= 9777.701 Pa

DOI: 10.1201/9781003481980-11

Example 11.3 A swimming pool is 4 m wide and 3 m deep. Calculate the force on the side of the pool.

```
w = 4 m;
UnitConvert[∫₀ₘ³ ᵐ ρ w g z dz, N]
```

Out[4]= 175 998.6 N

11.1.2 Air

Pressure is also measured in atmospheres (atm). Weather maps use the unit of bar which is defined as 10^5 Pa. Atmospheric pressure may be thought of as the weight of air molecules pushing on objects on the surface of the earth.

Example 11.4 Calculate the atm and bar in Pa.

```
In[5]:= UnitConvert[1. atm, Pa]
        UnitConvert[bar, Pa]
```

Out[5]= 101 325. Pa

Out[6]= 100 000 Pa

Air is about 1000 times less dense than water, and it is compressible. The density of the air depends on its temperature and pressure.

Example 11.5 Get the density of air.

```
In[8]:= ρ = ThermodynamicData["Air", "Density",
            {"Temperature" → Quantity[20, "DegreesCelsius"],
             "Pressure" → Quantity[1, "Atmospheres"]}]
```

Out[8]= 1.20458 kg/m³

Example 11.6 Suppose the density of air was constant at 1 atm. Estimate the thickness of the atmosphere.

```
In[9]:= Solve[ρ g h == 1 atm, h]
```

Out[9]= {{h → 8577.53 m}}

Example 11.6 gives the correct order of magnitude of the thickness of the atmosphere, however, the atmosphere is not uniform in density.

11.1.3 Law of Atmospheres

The atmosphere does not have a constant density because gravity is pulling the molecules toward the earth. At the same time, the molecules have kinetic energy and are trying to fly away, but they don't move fast enough, so it becomes exponentially harder for a molecule to reach an increasing height. Suppose the atmosphere is made up of molecules with mass m and number per volume n. The change in pressure (ΔP) from moving a vertical distance Δz is then

$$\Delta P = -mgn\Delta z.$$

Notice that the pressure is getting smaller with increasing z. Taking the limit where $\Delta z \to 0$, and taking z to be the height above ground, one gets

$$\frac{dP}{dz} = -mgn.$$

The ideal gas law (Chap. 12) gives

$$n = \frac{P}{kT},$$

so

$$\frac{dn}{dP} = \frac{1}{kT}.$$

Using the chain-rule for derivatives,

$$\frac{dn}{dz} = \frac{dn}{dP}\frac{dP}{dz} = -\frac{mg}{kT}n.$$

This equation has an easy solution because the exponential is the only function that gives itself back when differentiated.

Example 11.7 Solve for the number density of particles as a function of z.

In[10]:= $\mathsf{DSolve}\left[\left\{\mathsf{D[n[z], z]} == -\frac{\mathsf{m\ g}}{\mathsf{k\ T}}\mathsf{n[z]}, \mathsf{n[0]} == \mathsf{C}\right\}, \mathsf{n[z]},\right.$

$\left.\{\mathsf{z, 0, \infty}\}\right]$

Out[10]= $\left\{\left\{\mathsf{n[z]} \to \mathsf{C\ e}^{-\frac{\mathsf{g\ m\ z}}{\mathsf{k\ T}}}\right\}\right\}$

In Ex. 11.7, the number density at $z = 0$ has been set to a constant C.

Example 11.8 Use the exponential atmosphere to estimate its thickness as the distance by which the density drops by $1/e$.

In[11]:= m = [**nitrogen** CHEMICAL] [*molecular mass*] ; T = 300 K;

$$d = \text{UnitConvert}\left[\frac{k\ T}{m\ g}\right]$$

Out[12]= 9079.5 m

Example 11.9 Add up the weight of all the molecules in a vertical column of the atmosphere to calculate the pressure.

In[13]:= $\text{UnitConvert}\left[\rho\ g\ \int_{0\ m}^{10^{10}\ m} e^{-\frac{z}{d}}\ dz,\right.$

$\left.\text{Quantity}[1, \text{"Atmospheres"}]\right]$

Out[13]= 1.05852 atm

Adding up the weight of all the molecules, one finds the pressure to be about 1 atm. It is not exact because some approximations have been made for this estimate. Note that an isothermal atmosphere has been assumed. This is not quite true, the atmosphere gets colder at higher z. Also, the atmosphere was assumed to be only nitrogen.

11.2 PASCAL'S PRINCIPLE

Pascal's principle states that the pressure applied to a fluid appears everywhere in the fluid and the walls of the container. Consider a container (Figure 11.1) that has a cross-sectional area A_1 on the left and a_2 on the right. The pressure is the same on both sides at the top level of the fluid. Therefore,

$$\frac{F_1}{A_1} = \frac{F_2}{A_2},$$

where F_1 and F_2 are the forces on the left and right. Therefore a small force applied on the right can lift a large mass on the left (a hydraulic lift).

Figure 11.1 The pressure inside the container depends only on the height.

Example 11.10 A hydraulic lift is used to lift a 1800 kg car. What pressure is needed on small area side to lift the car if the large area is a circle of radius 22 cm?

In[128]:= m = 1800. kg; r = 22 cm; UnitConvert$\left[\dfrac{m\,g}{\pi\,r^2},\ \text{atm}\right]$

Out[128]= 1.14573 atm

Example 11.11 The top of a column of mercury is evacuated. To what height does the mercury rise?

In[130]:= ρ = [**mercury** ELEMENT] [*mass density*]; UnitConvert$\left[\dfrac{1\ \text{atm}}{\rho\,g}\right]$

Out[130]= 0.76343 m

11.3 ARCHIMEDES' PRINCIPLE

The expression for pressure in a fluid vs. depth h, $P = \rho g h$, can be evaluated for a cube at two different heights with the result that the difference in upward and downward forces on an object is equal to the weight of the fluid that is displaced by the object. This result is known as Archimedes' Principle. It holds for an object of any shape can be seen by replacing that object with the fluid that was displaced by that object.

Example 11.12 A string is attached to a copper ball of radius 1 cm. The ball is suspended from the string into water and the string is attached to a scale to measure its weight. What does the scale read in N?

In[27]:= ρ_w = [water CHEMICAL] [mass density] ; r = 1 cm;

$V = \frac{4}{3} \pi \ r^3$; ρ_c = 8.96 g/cm^3 ;

UnitConvert[ρ_c V g - ρ_w V g , N]

Out[29]= 0.327102 N

Weighing a object both in water and air allows determination of the density of the object because the buoyant force gives the volume of the object. Archimedes cleverly did this to determine if the king's crown was pure gold.

Example 11.13 The king's crown weights 1 kg in air and 0.948 kg in water. What is the density of the crown?

In[35]:= ρ_w = [water CHEMICAL] [mass density] ; W = 1 kg;

F_B = (1 - 0.9480) kg;

UnitConvert[$\dfrac{\frac{W}{g}}{\frac{F_B}{\rho_w \ g}}$, $\dfrac{g}{cm^3}$]

Out[37]= 19.174 g/cm^3

Example 11.14 Calculate the fraction of an iceberg that is exposed above water. Take the density of ice to be 0.917 g/cm^3 and the density of sea water to be 1.024 g/cm^3.

In[37]:= ρ_s = 1.024 $\dfrac{g}{cm^3}$; ρ_i = 0.917 $\dfrac{g}{cm^3}$; 1 - $\dfrac{\rho_i}{\rho_s}$

Out[37]= 0.104492

11.4 BERNOULLI'S EQUATION

Imagine fluid flowing in a pipe where the cross-sectional area of the pipe changes as well as its vertical height (Figure 11.2). The speed of the fluid must change according to conservation of energy.

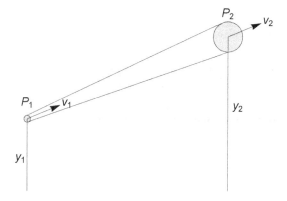

Figure 11.2 A fluid flows though a pipe at varying heights and speeds, resulting in varying pressure.

Bernoulli's equation reads

$$P_1 + \rho g y_1 + \frac{1}{2}\rho v_1^2 = P_2 + \rho g y_2 + \frac{1}{2}\rho v_2^2,$$

where P_1, y_1, v_1 and P_2, y_2, v_2 refer to pressure, height, and speed of the fluid in the pipe at two different positions.

Example 11.15 A fire extinguisher is discharged into the atmosphere and the fluid has an exit speed of 30 m/s. Calculate the pressure at the bottom of the tank if its height to the discharge point is 0.5 m.

In[17]:= **h = 0.5 m; v = 30 $\frac{m}{s}$;**

UnitConvert$\left[1 \text{ atm} + \rho_w \, g \, h + \frac{1}{2}\, \rho_w \, v^2, \text{ atm}\right]$

Out[18]= 5.47629 atm

CHAPTER 12

Thermodynamics

12.1 TEMPERATURE

Temperature is a measure of kinetic energy of a system of molecules. Temperature has meaning only in the context of thermal equilibrium. Two systems are in equilibrium if there is no net energy transfer between the two systems. Temperature (T) is measured on the absolute kelvin (K) scale. The conversion between degrees Celsius (T_C) and the absolute temperature is

$$T = \left(273.15 + \frac{T_C}{°C}\right)K.$$

A change of 1 °C is 1 K, but there is an offset of 273.15 K.

Example 12.1 Express 273.15 K in °C.

In[1]:= **UnitConvert$\left[$273.15 K, °C$\right]$**

Out[1]= 0. °C

In converting °C to K, one must be care of the format. Executing UnitConvert on 32 * °C gives (32) (273.15+1) K, while executing UnitConvert on (32 °C) gives the desired (273.15 +32) K. The first calculation is obtained by 32 followed by "°C" in the natural language box, while the second calculation is obtained by "32 °C" in the natural language box.

Example 12.2 Express 32 °C in K.

In[2]:= **UnitConvert$\left[$32. °C$\right]$**

Out[2]= 305.15 K

DOI: 10.1201/9781003481980-12

Example 12.3 Express 273.15 K in °F.

In[3]:= **UnitConvert[273.15 K, °F]**

Out[3]= 32. °F

The conversion between temperature on the Fahrenheit scale (T_F) is

$$T_F = \frac{9°F}{5°C} T_c + 32°F.$$

Example 12.4 Express 100 °C in °F.

In[4]:= **UnitConvert[100. °C, °F]**

Out[4]= 212. °F

Example 12.5 Express 100 °F in K.

In[5]:= **UnitConvert[100. °F]**

Out[5]= 310.928 K

12.2 HEAT

Heat is energy that is transferred from an object at a lower temperature to an object at a higher temperature. It is often represented by the symbol Q. The heat may be transferred in three ways: (1) direct contact (conduction), (2) moving fluids (convection), and (3) radiation.

12.2.1 Conduction

Consider conduction of heat through a wall. The rate is

$$\frac{\Delta Q}{\Delta t} = -\kappa A \frac{\Delta T}{\Delta x},$$

where A and Δx are the area and thickness of the wall and κ is the thermal conductivity.

Example 12.6 A foam insulator has a thermal conductivity of 0.024 W/(m K). Calculate the rate of energy loss through a 10 ft by 10 ft wall that is 6 in thick when the temperature difference is 20 K.

$$\text{In[7]:= } \kappa = 0.024 \ \frac{W}{m \ K}; \ A = 100 \ ft^2; \ \Delta T = 20 \ K; \ \Delta x = 6 \ in; \ -\kappa \ A \ \frac{\Delta T}{\Delta x}$$

Out[7]= -29.2608 W

The R value is commonly used in construction. It is defined as some thickness t divided by the thermal conductivity.

$$R = \frac{t}{\kappa}.$$

Typically, this is calculated per inch of insulator ($t = 1$ in) and given in units of (ft^2 °F h) per BTU.

Example 12.7 Calculate the R value for 1 inch of material with a thermal conductivity of 0.024 W/(m K).

$$\text{In[8]:= } \kappa = 0.024 \ \frac{W}{m \ K};$$

$$R = \text{UnitConvert}\left[\frac{1 \ in}{\kappa}, \ \frac{ft^2 \ °F \ h}{BTU_{IT}}\right];$$

$$\frac{R \ BTU_{IT}}{K \ ft^2 \ h} \ 1.8 \ "BTU \ per \ h \ per \ ft^2"$$

Out[9]= 6.0095 BTU per h per ft^2

12.2.2 Convection

Heat is transported by convection, which is the mass in motion. This motion is often turbulent, and details of the convection may be difficult to calculate. Simple scenarios may be described by Newton's law of cooling,

$$\frac{\Delta Q}{\Delta t} = hA(T_s - T_f).$$

where h is the convective heat transfer coefficient, T_s is the surface temperature, and T_f is the temperature of the moving fluid. The above equation gives the rate of heat transfer from the surface to the fluid.

Example 12.8 Consider cool air at $T = 20°C$ flowing over a flat hot surface of area 10 m² that is maintained at $T = 80°C$. The heat transfer coefficient is $h = 75$ W/(m² K). Calculate the convection rate.

In[10]:= T_s = 80. K; T_f = 20. K; A = 10. m²; h = 75. $\frac{W}{m^2 \, K}$;

h A (T_s - T_f)

Out[11]= 45 000. W

12.2.3 Radiation

The formula for the total power radiated comes from modern physics. The power per area (P/A) is proportional to the 4th power of the temperature. The constant of proportionality is called the Stefan-Boltzmann constant. The radiation result is written

$$\frac{P}{A} = \sigma T^4.$$

Example 12.9 Get the Stefan-Boltzmann constant.

In[12]:= **Quantity["StefanBoltzmannConstant"]**
      UnitConvert[1. × %]

Out[12]= σ

Out[13]= 5.67037×10^{-8} kg/($s^3 K^4$)

12.3 BOLTZMANN CONSTANT

The Boltzmann constant (Ex. 1.26) converts absolute temperature to energy. A convenient way to remember the conversion factor is that at room temperature

$$kT \approx \frac{1}{40} \, eV.$$

Example 12.10 Get kT at room temperature.

In[14]:= **T = 290. K; Rationalize[UnitConvert[k T, eV], 10^{-2}]**

Out[14]= $\dfrac{1}{40}$ eV

12.4 IDEAL GAS LAW

The ideal gas law relates the number of molecules, volume, pressure, and temperature. It reads

$$PV = NkT.$$

Standard temperature and pressure (STP) is $T = 272.15$ K and $P = 1$ atm.

Example 12.11 Calculate the volume of 1 mol of gas molecules at STP.

In[15]:= **P = 1 atm; T = 273.15 K; N = N_Θ; UnitConvert$\left[\dfrac{N\ k\ T}{P}\right]$**

Out[15]= $0.022414\,\mathrm{m}^3$

Example 12.12 Calculate the number of molecules in 2 liters of gas at a temperature of 300 K and a pressure of 1 atm. How many moles is this?

In[16]:= **T = 300 K; V = 2. L; P = 4 atm;**

N = UnitConvert$\left[\dfrac{P\,V}{k\ T}\right]$

UnitConvert$\left[\dfrac{N}{\frac{N_\Theta}{\mathrm{mol}}}\right]$

Out[16]= 1.95705×10^{23}

Out[17]= $0.324976\,\mathrm{mol}$

An alternate form of the gas law expresses the number of particles N in moles. The gas constant R is defined by

$$N_0 k = (1 \text{ mol})R,$$

where N_0 is Avogadro's number. Thus, the gas law may be written

$$PV = nRT,$$

where n is the number of moles.

Example 12.13 Verify that $N_0 k = (1 \text{ mol})R$.

```
In[18]:=  N₀ k == (1 mol) R
```

```
Out[18]= True
```

A pressure vs. volume plot is a useful way to analyze a gas (see Sect. 12.6). One must keep in mind that P and V are not the only variables because T (or even the number of particles) can vary. Thus, one has to know what happens to T in order to make the plot. One simple case is that T remains constant. Then P varies as V^{-1}. Figure 12.1 shows a plot of P vs. V at three different temperatures. The upper (lower) curve is the highest (lowest) temperature.

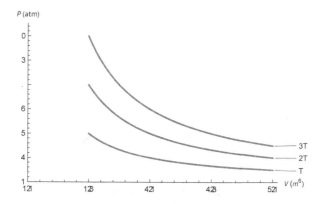

Figure 12.1   The temperature of a gas remains constant. The plot of P vs. V is shown for three different temperatures.

12.5 HEAT CAPACITY

Heat capacity C is given by the amount of energy Q needed to raise the temperature by ΔT,

$$C = \frac{Q}{\Delta T}.$$

Normalized to mass (m), this quantity is called the specific heat (c),

$$c = \frac{Q}{m\Delta T}.$$

It is usually calculated per mole of substance.

For a process occurring at constant volume, the molar heat capacity C_V is

$$C_V = \frac{\Delta Q}{n\Delta T}.$$

The equipartition theorem states that the average molecular energy is $\frac{1}{2}kT$ times the number of degrees of freedom. For a monatomic gas there are 3 translational degrees of freedom giving $U = \frac{3}{2}kT$. A diatomic molecule may also have 2 rotational degrees of freedom giving $U = \frac{5}{2}kT$, and in addition, 2 vibrational degrees of freedom giving $U = \frac{7}{2}kT$.

The heat capacity at constant volume for an ideal gas is

$$C_V = \frac{3}{2}R.$$

For a diatomic ideal gas, the typical value is

$$C_V = \frac{5}{2}R.$$

12.6 FIRST LAW

The first law of thermodynamics is a statement of energy conservation. If you add heat (Q) to a system, it can either do work (W) or cause a change in internal energy (ΔU), or both. The first law reads

$$Q = W + \Delta U.$$

It is useful to look to the first law for an ideal gas under various conditions. An expanding gas does work according to

$$W = \int_{V_1}^{V_2} dV\, P = Nk \int_{V_1}^{V_2} dV\, \frac{T}{V}.$$

Keep in mind that T may be changing as the gas expands. In a plot of P vs. V, the work done by the gas is the area under the curve.

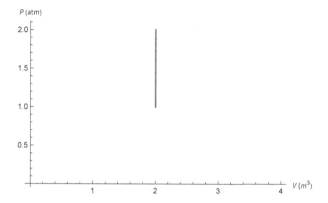

Figure 12.2 A gas that has constant volume does zero work. The area under the curve in a plot of P vs. V is zero.

12.6.1 Constant Volume

A process at constant volume is referred to as an isometric process. For constant volume (Figure 12.4), zero work is done, and the first law reads

$$Q = \Delta U.$$

If the pressure is increasing (or decreasing), then the temperature is increasing (or decreasing) and heat is being added (or subtracted).

The heat capacity at constant volume is defined by

$$Q = nC_V\Delta T,$$

Since

$$\Delta U = nC_V\Delta T,$$

the heat added is

$$Q = \Delta U = nC_V\Delta T.$$

12.6.2 Constant Pressure

A process at constant pressure is referred to as an isobaric process. For constant pressure P, the work done is the pressure times the change in volume (ΔV),

$$W = P\Delta V.$$

An increase (decrease) in volume means that work done by the gas is positive (negative). Negative work done on the gas means that work is done by the gas.

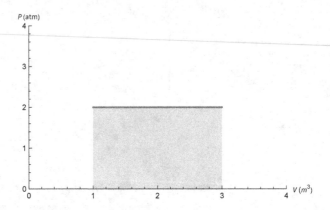

Figure 12.3 A gas that has constant pressure does positive (negative) work if it expands (contracts). The area under the curve in a plot of P vs. V is the work.

The first law reads

$$Q = P\Delta V + \Delta U.$$

If the work is positive ($\Delta V > 0$), then the temperature has increased and heat has been added ($Q > 0$). If the work is negative ($\Delta V < 0$), then the temperature has decreased and heat has been subtracted from the gas ($Q < 0$).

Example 12.14 A gas with initial volume 1 m³ and pressure 2 atm expands at constant pressure to a volume of 3 m³. Calculate the work done by the gas.

In[18]:= **Clear[V]; W =** $\int_{1\ m^3}^{3\ m^3}$ **2. atm d V;**

UnitConvert[W, J]

Out[19]= 405 300. J

The heat capacity at constant pressure is defined to be

$$Q = nC_P\Delta T.$$

The first law reads

$$Q = P\Delta V + nC_V\Delta T.$$

The gas law gives

$$P\Delta V = nR\Delta T.$$

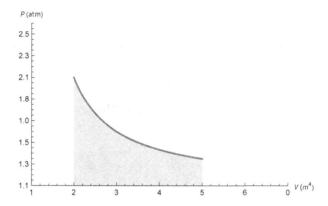

Figure 12.4 A gas that has constant temperature does positive (negative) work if it expands (contracts). The area under the curve in a plot of P vs. V is the work.

Combining the expressions gives

$$nC + P\Delta T = nR\Delta T + nC_V\Delta T.,$$

or

$$C_P = C_V + R.$$

12.6.3 Constant Temperature

A process at constant temperature is referred to as an isothermal process. The work done by the gas is

$$W = \int_{V_1}^{V_2} dV \, P = NkT \int_{V_1}^{V_2} dV \, \frac{1}{V} = NkT \ln \frac{V_2}{V_1}.$$

Example 12.15 A gas with initial volume 1 m^3 and pressure 1 atm expands isothermally to a volume of 4 m^3. Calculate the work done by the gas.

```
In[20]:= W = ∫₁ ₘ³⁴ ᵐ³  1. atm m³ / V  dV;

        UnitConvert[W, J]

Out[21]= 140 466. J
```

12.6.4 Adiabatic

If there is a sudden expansion or contraction of a gas, there is no time for heat to be added or escape, and

$$Q = 0.$$

Such a process is called an adiabatic process. The first law reads

$$0 = W + \Delta U.$$

For an ideal gas,

$$PV = nRT,$$

and taking the differential of each side,

$$PdV + VdP = nRdT.$$

From the first law,

$$0 = PdV + nC_V dT,$$

and combining the two expressions gives

$$0 = PdV + nC_V \frac{PdV + VdP}{nR} = PdV + C_V \frac{PdV + VdP}{R}.$$

Multiplying by R gives

$$0 = RPdV + C_V PdV + C_V VdP = (C_C + R)PdV + C_V VdP.$$

Using the expression for heat capacity at constant pressure (Sect. 12.14), $C_P = C_V + R$, one gets

$$0 = C_P PdV + C_V VdP.$$

Therefore,

$$\frac{dP}{P} + \frac{C_P}{C_V} \frac{dV}{V} = 0.$$

This result is

$$\ln\left(PV^{C_P/C_V}\right) = \text{constant},$$

or conventionally written,

$$PV^{\gamma} = \text{constant},$$

where $\gamma = \frac{C_P}{C_V}$. Since $\gamma > 1$, the pressure in an adiabatic expansion (contraction) drops (rises) faster than an isothermal expansion (contraction).

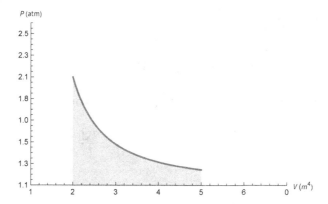

Figure 12.5 A gas that has no heat transfer (PV^γ = constant) does positive (negative) work if it expands (contracts). The area under the curve in a plot of P vs. V is the work.

For a monatomic ideal gas,

$$\gamma = \frac{\frac{3}{2}R + R}{\frac{3}{2}R} = \frac{5}{3}.$$

For a diatomic ideal gas, the typical value is

$$\gamma = \frac{\frac{5}{2}R + R}{\frac{5}{2}R} = \frac{7}{5}.$$

Example 12.16 A gas with initial volume 1 m³ and pressure 1 atm expands adiabatically to a volume of 4 m³. Calculate the work done by the gas.

```
In[22]:= W = ∫₁ m³⁴ m³ (1. atm (m³)^(7/5))/(V^(7/5)) dV;

        UnitConvert[W, J]
```

```
Out[23]= 107 823. J
```

Note that in an adiabatic expansion, $P \sim V^{-\gamma}$ with $\gamma > 1$, and the pressure drops faster than an isothermal expansion $P \sim V^{-1}$. Compare Figs. 12.4 and 12.5.

12.7 ENTROPY

Entropy (S) is defined by

$$dS = \frac{dQ}{T}.$$

The units of entropy are J/K. The change in entropy in going from state A to state B is

$$\Delta S = S(B) - S(A) = \int_A^B \frac{dQ}{T}.$$

For an isothermal process,

$$\Delta S = \frac{1}{T} \int dQ = \frac{1}{T} \int dW = \frac{1}{T} \int dV p = Nk \int \frac{dV}{V}.$$

For an expansion from V_i to V_f this gives

$$\Delta S = Nk \ln \frac{V_f}{V_i} = Nk \ln \frac{p_i}{p_f}.$$

Example 12.17 Calculate the change in entropy when 1 mol of an ideal gas has an isothermal expansion from an initial pressure of 1 atm to a final pressure of 0.5 atm.

In[24]:= **UnitConvert** $\left[N_{\theta} \ k \ \text{Log} \left[\dfrac{1 \ \text{atm}}{0.5 \ \text{atm}} \right], \dfrac{J}{K} \right]$

Out[24]= 5.76315 J/K

Example 12.18 Two ends of a metal rod are connected to thermal reservoirs at 1500 K and 300 K and the flow of heat through the rod is 100 J/s. Calculate the rate of change of entropy.

In[25]:= **UnitConvert** $\left[\dfrac{100 \ J}{s} \left(\dfrac{1}{300. \ K} - \dfrac{1}{1500. \ K} \right), \dfrac{J}{K \ s} \right]$

Out[25]= 0.266667 J/(s K)

Example 12.19 Find the change in entropy when 10 lb of ice melts. It takes 334 J to melt 1 g of ice.

In[26]:= **10. lb** $\dfrac{334. \ J}{1 \ g} \dfrac{1}{273 \ K}$

Out[26]= 5549.45 J/K

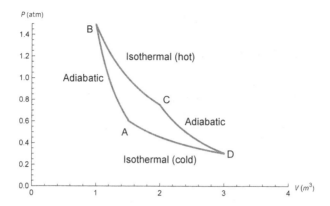

Figure 12.6 The Carnot cycle consists of an adiabatic compression (A → B), an isothermal expansion (B → C), an adiabatic expansion (C → D), and an adiabatic compression (D → A).

12.8 SECOND LAW

The second law states that the total entropy cannot decrease for any process. The best it can do is not change. This is usually formulated in terms of an engine efficiency: It is not possible to construct a cyclic engine that converts thermal energy into work with 100% efficiency. A cyclic cycle may be conveniently depicted on a diagram of p vs. V. The process that produces the maximum efficiency is called the Carnot cycle (Figure 12.6). The Carnot engine operates between hot and cold reservoirs. It consists of an adiabatic compression, an isothermal expansion which does positive work, an adiabatic expansion that discharges heat, and an isothermal compression that does negative work.

The Carnot cycle adds heat (Q_h) at the high temperature and discharges heat (Q_c) at the cold temperature. For a complete cycle, there is no change in internal energy so the net heat added is equal to the work done by the system

$$Q_h - Q_c = W.$$

The efficiency ε of the engine is

$$\varepsilon = \frac{W}{Q_h} = \frac{Q_h - Q_c}{Q_h} = 1 - \frac{Q_c}{Q_h}.$$

Example 12.20 Calculate the efficiency of the Carnot cycle in terms of the hot and cold temperatures.

In[27]:= `ClearAll["Global`*"];`

$$W_{BC} = n\,R\,T_h\,\mathbf{Log}\!\left[\frac{V_C}{V_B}\right]; \quad W_{DA} = n\,R\,T_c\,\mathbf{Log}\!\left[\frac{V_A}{V_D}\right];$$

$$p_C = \frac{p_D\,V_D{}^\gamma}{p_A\,V_A{}^\gamma}\,\frac{p_B\,V_B{}^\gamma}{V_C{}^\gamma}; \quad p_C = \frac{V_B}{V_C}\,p_B; \quad V_C = \frac{V_D}{V_A}\,V_B;$$

$$\varepsilon = 1 - \frac{-W_{DA}}{W_{BC}} \;/.\; \mathbf{Log}\!\left[\frac{V_D}{V_A}\right] \rightarrow -\mathbf{Log}\!\left[\frac{V_A}{V_D}\right]$$

Out[30]= $1 - \dfrac{T_c}{T_h}$

CHAPTER 13

Waves

13.1 WAVE EQUATION

Consider a wave traveling in the z-direction with speed v. The wave may be described by a function $y(z,t) = y(z - vt)$. The wave equation relates two space derivatives with two time derivatives

$$\frac{\partial^2 y}{\partial z^2} = \frac{1}{v^2} \frac{\partial^2 y}{\partial t^2},$$

which are connected by the wave speed.

13.2 TRAVELING WAVES

The traveling wave may be written

$$y(z,t) = A\cos(kz - \omega t).$$

Figure 13.1 shows a traveling wave.

Since the argument of the cosine function is dimensionless, k must have units of inverse m. When z changes by a quantity called the wavelength λ with t fixed, then y does not change. The periodicity of the cosine function gives

$$k = \frac{2\pi}{\lambda}.$$

The quantity k is called the wave number.

DOI: 10.1201/9781003481980-13

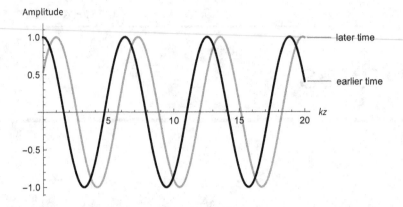

Figure 13.1   The amplitude of a traveling wave is shown as a function of kz at time $t = 0$, then again at a slightly later time.

Example 13.1 Show that y does not change when $z \to z + \lambda$.

In[4]:= $k = \dfrac{2\,\pi}{\lambda}$; Cos[k z - ω t] /. {z → z + λ} // Simplify

Out[4]= $\text{Cos}\left[\dfrac{2\,\pi\,z}{\lambda} - t\,\omega\right]$

Similarly, ω must have units of s^{-1}. When t changes by a quantity called the period T with z fixed, then f does not change. The periodicity of the cosine function gives

$$\omega = \frac{2\pi}{T}.$$

Example 13.2 Show that y does not change when $t \to t + T$.

In[9]:= Clear[k]; $\omega = \dfrac{2\,\pi}{T}$; Cos[k z - ω t] /. {t → t + T} // Simplify

Out[9]= $\text{Cos}\left[\dfrac{2\,\pi\,t}{T} - k\,z\right]$

The wave equation gives

$$v = \frac{\omega}{k}.$$

Example 13.3 Use the wave equation to find an expression for v.

```
In[10]:= $Assumptions = {k > 0, ω > 0, v > 0};
         f = Cos[k z - ω t];
         Solve[∂z ∂z f == 1/v² ∂t ∂t f, v]
```

$$\text{Out[10]}= \left\{\left\{v \to \frac{\omega}{k}\right\}\right\}$$

In summary, one may write

$$v = \frac{\omega}{k} = \frac{\omega}{2\pi}\lambda = \frac{\lambda}{T} = \lambda f.$$

The time average of the wave comes from the fundamental relationship between sine and cosine,

$$\sin^2\theta + \cos^2\theta = 1.$$

Example 13.4 Calculate the time average of the function $\cos\left(kz - \frac{2\pi t}{T} + \delta\right)$ over one period T.

```
In[14]:= 1/T ∫₀ᵀ Cos[k z - 2 π t / T + δ]² dt
```

$$\text{Out[14]}= \frac{1}{2}$$

13.3   GROUP AND PHASE VELOCITY

One may form a "group" of 2 waves with different wave numbers and frequencies by super position,

$$y(z,t) = A\cos(k_1 z - \omega_1 t) + A\cos(k_2 z - \omega_2 t).$$

We can use trigonometric identities for the cosine of the sum and difference of two quantities to write the wave group in a form that we can visualize.

Example 13.5 Expand $\cos(a+b)$.

```
In[5]:= Expand[Cos[a + b], Trig → True]
```

Out[5]= Cos[a] Cos[b] - Sin[a] Sin[b]

Example 13.6 Expand $\cos(a - b)$.

In[6]:= **Expand[Cos[a - b], Trig → True]**

Out[6]= **Cos[a] Cos[b] + Sin[a] Sin[b]**

From these identities, one may see that

$$\cos a + \cos b = 2\cos\left(\frac{a+b}{2}\right)\cos\left(\frac{a-b}{2}\right).$$

Example 13.7 Show the above expression for $\cos a + \cos b$ is valid.

In[7]:= **2 Cos$\left[\dfrac{a + b}{2}\right]$ Cos$\left[\dfrac{a - b}{2}\right]$ // Simplify**

Out[7]= **Cos[a] + Cos[b]**

Thus,

$$y(z,t) = 2A\cos\left(\frac{k_1 z - \omega_1 t + k_2 z - \omega_2 t}{2}\right)\cos\left(\frac{k_1 z - \omega_1 t - k_2 z + \omega_2 t}{2}\right),$$

or

$$y(z,t) = 2A\cos\left(\frac{k_1 + k_2}{2}z - \frac{\omega_1 + \omega_2}{2}t\right)\cos\left(\frac{k_1 - k_2}{2}z - \frac{\omega_1 - \omega_2}{2}t\right),$$

Now consider the case where ω_1 and ω_2 are approximately equal. In this case, one of the cosines (the sum) has a fast oscillation and one (the difference) has a slow oscillation. This is shown in Figure 13.2.

Consider the second cosine as a time-dependent amplitude $B(t)$ of the fast oscillations, that changes slowly with time,

$$y(z,t) = B(t)\cos\left(\frac{k_1 + k_2}{2}z - \frac{\omega_1 + \omega_2}{2}t\right).$$

The fast oscillation corresponds to the phase velocity and the slow oscillation to the group velocity.

13.4   STANDING WAVES

The traveling wave of Figure 13.1 may be thought of as an oscillation of each point in space (z) oscillating up and down with time. Now imagine that wave

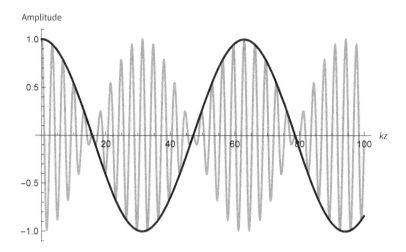

Amplitude

Figure 13.2 Two waves with slightly different frequencies are added together. This results in a fast oscillation with slowly varying amplitude. The varying amplitude of the fast oscillations is described by a slow oscillation.

confined to fit into some interval of z. Standing waves occur when an exact number of half-wavelengths fit into the specified interval. Figure 13.3 shows a standing wave in which 1/2 wavelength fits in the interval. At every point in the interval, the amplitude of the wave oscillates synchronously between the two curves. There are nodes at each end of the interval.

A standing wave is constructed by adding two traveling waves that are moving in opposite directions. Recall that the algebraic sign inside the trigonometric function gives the direction of wave travel. For travel in the $+z$ direction,

$$y_1(z,t) = A\sin(kz - \omega t),$$

and for travel in the $-z$ direction,

$$y_2(z,t) = A\sin(kz + \omega t).$$

Adding them together gives

$$y_1 + y_2 = 2A\sin(kx)\cos(\omega t).$$

The addition of the two waves traveling in opposite direction is no longer a traveling wave as it does not have the proper $z - vt$ or $z + vt$ form.

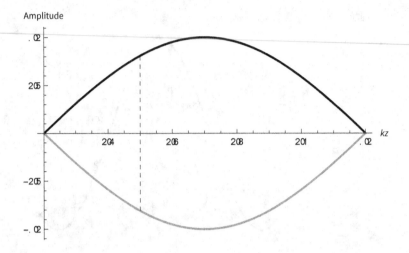

Figure 13.3 The boundaries of the amplitude are shown for a standing wave in which 1/2 wavelength fits in a specified interval. The amplitude at every point oscillates between the two curves (dashed line). There are nodes on each end where the amplitude is always zero.

Example 13.8 Add the two waves traveling in opposite directions.

```
In[8]:= ClearAll["Global`*"];
        Expand[A Sin[k z - ω t] + A Sin[k z + ω t], Trig → True]

Out[8]= 2 A Cos[t ω] Sin[k z]
```

Figure 13.5 shows standing wave in which 1, 3/2, 2, and 5/2 wavelength fits the interval. Additional nodes appear.

13.5 WAVE ON A STRING

Consider a pulse on a string. The force causing the centripetal acceleration is the tension in the string (T).

$$F = 2T \sin\theta = \frac{T \Delta s}{r} = \frac{\Delta m v^2}{r}.$$

Solving for v,

$$v = \sqrt{\frac{T \Delta s}{\Delta m}} = \sqrt{\frac{T}{\mu}},$$

where $\mu = \Delta m / \Delta s$.

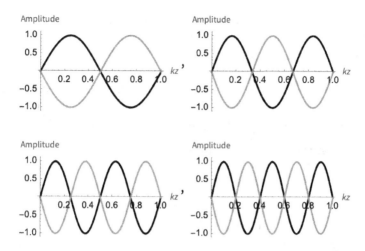

Figure 13.4 Standing waves are shown for 1, 3/2, 2, and 5/2 wavelength fit to the interval.

Example 13.9 A string has a density of 0.25 kg/m. Calculate the speed of a wave on the string when the tension is 100 N.

In[9]:= μ = 0.25 $\dfrac{\text{kg}}{\text{m}}$; T = 100 N ; v = UnitConvert$\left[\sqrt{\dfrac{T}{\mu}} \right]$

Out[9]= 20. m/s

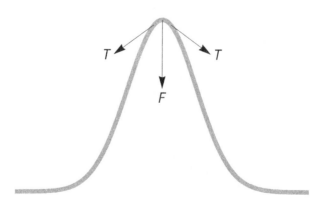

Figure 13.5 For a pulse on a string, the downward force on a tiny piece of the string is caused by the tension.

Notice that the v calculated in Ex. 13.9 is the horizontal velocity of the propagation of the wave. If the wave is sinusoidal, then the transverse motion of each piece of the string is simple harmonic motion.

Example 13.10 For the string of Ex. 13.9, a sinusoidal wave is propagated with a frequency of 150 Hz and amplitude 0.9 cm. Calculate the transverse velocity and acceleration of the string at $x = 0.25$ m and $t = 2$ s.

```
In[10]:= f = 150 s⁻¹; A = 0.009 m;
              v          2 π
         λ = ─; k = ───; ω = v k;
              f          λ
         y = A Sin[k x - ω t] // Simplify;
         ∂t y /. {x → 0.25 m, t → 2 s}
         ∂t ∂t y /. {x → 0.25 m, t → 2 s}
```

```
Out[12]= -5.99789 m/s
```

```
Out[13]= 5652.88 m/s²
```

13.6 POWER

The power P transmitted by a wave on a string is the transverse velocity of the string times the component of the tension in that direction,

$$P = \frac{\partial y}{\partial t} T_y.$$

The component of tension T_y is the tension times the slope of the string,

$$T_y = T \frac{\partial y}{\partial x},$$

which gives

$$P = \frac{\partial y}{\partial t} T \frac{\partial y}{\partial x}.$$

Averaging the power over time gives a factor of 1/2.

Example 13.11 Calculate the average transmitted power for a sinusoidal wave on a string.

In[14]:= **ClearAll["Global`*"]; y = A Sin[k x - ω t];**

P = -∂$_t$ y T ∂$_x$ y;

$$\frac{\omega}{2\,\pi}\ \textbf{Integrate}\left[\textbf{P, }\left\{\textbf{t, 0, }\frac{2\,\pi}{\omega}\right\}\right]\ \textbf{/.}\ \left\{\textbf{T} \rightarrow \mu\ \textbf{v}^2\textbf{, }\textbf{k} \rightarrow \frac{\omega}{\textbf{v}}\right\}$$

Out[16]= $\dfrac{1}{2}\ A^2\ v\,\mu\,\omega^2$

Example 13.12 Give a numerical value of the transmitted power for the wave described in Ex. 13.9 and Ex. 13.10.

In[17]:= **v = 20 $\dfrac{m}{s}$; μ = 0.25 $\dfrac{kg}{m}$; f = 150 s^{-1};**

A = 0.009 m; λ = $\dfrac{v}{f}$; k = $\dfrac{2\,\pi}{\lambda}$; ω = v k;

UnitConvert$\left[\dfrac{1}{2}\ A^2\ v\ \mu\ \omega^2\textbf{, }W\right]$

Out[19]= 179.874 W

13.7   SOUND WAVES

13.7.1   Speed of Sound Waves

Sound waves are longitudinal waves of compression and rarefaction of a fluid. Consider the oscillation of pressure in a tube of cross-sectional area S. Newton's law applied to the pressure difference over a time Δt is

$$F = pS - (p + \Delta p)S = -\Delta pS = ma = \rho vS\,\Delta t\frac{\Delta v}{\Delta t},$$

One can write the change in volume over volume as

$$\frac{\Delta V}{V} = \frac{S\,\Delta v\Delta t}{S\,v\Delta t} = \frac{\Delta v}{v},$$

to arrive at

$$v = \sqrt{\frac{\frac{-\Delta p}{\Delta V/V}}{\rho}}.$$

The quantity in the numerator of the square root is called the bulk modulus (B) and one writes the speed of the sound wave as

$$v = \sqrt{\frac{B}{\rho}}.$$

Example 13.13 The bulk modulus of air is 142 kPa. Calculate the speed of sound at 20°C and 1 atm.

In[20]:= **B = 142 10³ Pa;**

ρ = **ThermodynamicData["Air", "Density",**

{"Temperature" → 20 °C, "Pressure" → 1 atm}];

UnitConvert$\left[\sqrt{\dfrac{B}{\rho}}\right]$

Out[21]= 343.342 m/s

Water is not nearly as compressible as air, and thus, has a higher speed of sound.

Example 13.14 The bulk modulus of water is 2.2 GPa. Calculate the speed of sound.

In[22]:= **B = 2.2 10⁹ Pa;** ρ = [**water** CHEMICAL][*mass density*];

UnitConvert$\left[\sqrt{\dfrac{B}{\rho}}\right]$

Out[23]= 1485.43 m/s

13.7.2 Wave Form

The longitudinal displacement $s(x,t)$ of the sound wave is

$$s = A\cos(kx - \omega t).$$

This may be thought of as a pressure wave with

$$\Delta p = -B\frac{\Delta V}{V} = -B\frac{S\,\Delta s}{S\,\Delta x}.$$

In the limit where $\Delta x \to 0$,

$$\Delta p = -B\frac{\partial s}{\partial x} = BkA\sin(kx - \omega t) = k\rho v^2 A\sin(kx - \omega t).$$

Note that the pressure is out of phase with the displacement by $\pi/2$ (cosine vs. sine).

Example 13.15 The faintest sound that can be heard at 1000 Hz has an amplitude 0.011 nm. Take the density of air to be 1.21 kg/m^3 and the speed of sound to be 343 m/s. Calculate the pressure amplitude.

In[24]:= **A = 0.011 nm ; f = 1000 Hz ;**

$$\mathbf{k = \frac{2\,\pi\,f}{v}\,;\; v = 343\;\frac{m}{s}\,;\; \rho = 1.1\;\frac{kg}{m^3}\,;}$$

UnitConvert$\left[k\,\rho\,v^2\,A,\;Pa\right]$

Out[26]= 0.000447222 Pa

13.7.3 Intensity

The power (P) transmitted by the fluid is the oscillation speed ($\partial s/\partial t$) times the force,

$$P = \frac{\partial s}{\partial t}F = \omega A\sin(kx - \omega t)F.$$

In terms of the pressure amplitude ($A_p = k\rho v^2 A$), the force is

$$F = SA_p\sin(kx - \omega t).$$

The power may be written, using $v = \omega/k$, as

$$P = \frac{SA_p^2}{\rho v}\sin^2(kx - \omega t).$$

The intensity I is the time averaged power (P_{ave}) per area. Since the time average of the sine squared is 1/2, the intensity is

$$I = \frac{P_{ave}}{S} = \frac{A_p^2}{2\rho v}.$$

Sound intensity is measured on a logarithmic scale in units of decibel (db) defined by a dimensionless quantity β,

$$\beta = 10\log_{10}\frac{I}{I_0},$$

where $I_0 = 10^{-12}$ W/m^2.

Example 13.16 A sound source has a power output of 100 W. The sound waves travel in all directions. Calculate the intensity in db 8 m from the source.

In[27]:= **P = 100 W; r = 8. m; I = $\dfrac{P}{4 \pi r^2}$; I$_\theta$ = 10^{-12} $\dfrac{W}{m^2}$;**

10 Log$\left[10, \dfrac{I}{I_\theta}\right]$ "db"

Out[28]= **110.946 db**

13.7.4 Doppler Effect

Consider a source that is moving toward the observer with speed v_s (Figure 13.6). The speed does not change, but the wavelength gets shorter and the frequency gets larger. Let f' and λ' be the frequency and wavelength of the source, and f and λ be the frequency and wavelength received by the observer. The frequency heard by the observer is

$$f' = \frac{v}{\lambda'} = \frac{v}{\frac{v_s-v}{f}} = f\frac{v}{v_s-v} = \frac{f}{1-\frac{v_s}{v}},$$

and the wavelength is

$$\lambda' = \frac{v-v_s}{f}.$$

If the source moves away from the observer, then the sign of v_s is reversed.

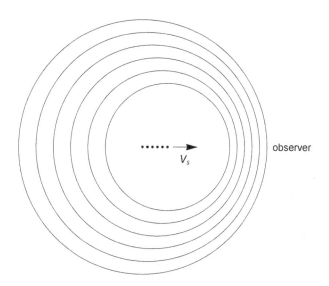

observer

Figure 13.6 Wavefronts for a moving source are shown at equally spaced time intervals. An observer for which the source is approaching has the wavefronts closer spaced, corresponding to shorter wavelength and higher frequency.

Example 13.17 A car moves toward the observer at a speed of 60 mi/h, sounding a horn at 800 Hz. What frequency is heard when the car moves toward and away from a stationary observer?

In[29]:= $v = 343 \frac{m}{s}$; $v_s = 60. \frac{mi}{h}$; $f = 800$ Hz ;

$$\frac{f}{1 - \frac{v_s}{v}}$$

$$\frac{f}{1 + \frac{v_s}{v}}$$

Out[30]= 867.867 Hz

Out[31]= 741.978 Hz

Now consider an observer moving with speed v_o toward a stationary source. The observer now measures an increased wave speed,

$$v' = v + v_o.$$

The observed frequency is

$$f' = \frac{v'}{\lambda} = \frac{v + v_0}{\lambda} = f\left(1 + \frac{v_0}{v}\right),$$

and the wavelength is unchanged.

If the observer moves away from the source, then the sign of v_0 is reversed.

Example 13.18 A moving observer hears a frequency of 800 Hz (700 Hz) when moving toward (away) from a stationary sound source? Determine the speed of the observer and the frequency of the source.

In[32]:= **<< Notation`**

    Symbolize[ParsedBoxWrapper[SubscriptBox["_", "_"]]]

In[34]:= **ClearAll["Global`*"];**

    sol = Solve$\left[f_{\prime_+} == f\left(1 + \frac{v_0}{v}\right) \;\&\&\; f_{\prime_-} == f\left(1 - \frac{v_0}{v}\right), \{f, v_0\}\right]$ //

    Simplify

Out[34]= $\left\{\left\{f \rightarrow \dfrac{1}{2}\,(f_{\prime_-} + f_{\prime_+}),\; v_0 \rightarrow \dfrac{(-f_{\prime_-} + f_{\prime_+})\,v}{f_{\prime_-} + f_{\prime_+}}\right\}\right\}$

In[35]:= **f /. sol[[1]] /. $\{f_{\prime_+} \rightarrow 800\ \text{Hz},\; f_{\prime_-} \rightarrow 700\ \text{Hz}\}$**

    v_0 /. sol[[1]] /. $\left\{f_{\prime_+} \rightarrow 800\ \text{Hz},\; f_{\prime_-} \rightarrow 700\ \text{Hz},\; v \rightarrow 343.\ \dfrac{\text{m}}{\text{s}}\right\}$

Out[35]= **750 Hz**

Out[36]= **22.8667 m/s**

If both the source and observer are moving, then

$$\lambda' = \frac{v - v_s}{f},$$

$$v' = v - v_0,$$

and

$$f' = \frac{v'}{\lambda'} = f\frac{v - v_0}{v - v_s}.$$

There are four choices here because v_s and v_o can be either positive or negative relative to the direction of the speed of sound v.

Example 13.19 An observer and the sound source are moving toward each other. The observer has a speed of 30 m/s and the source has a speed of 25 m/s. The frequency if the source is 800 Hz. What frequency does the observer hear?

```
In[37]:= ClearAll["Global`*"];

        v₀ = -30 m/s ; vₛ = 25 m/s ; v = 343. m/s ; f = 800 Hz ;

        f (v - v₀)/(v - vₛ)
```

Out[39]= 938.365 Hz

13.8 LIGHT WAVES

Light waves are oscillations of electric and magnetic fields. They can propagate in a vacuum because changing electric and magnetic fields generate the fields. The speed of light is denoted by the symbol c.

Example 13.20 Get the speed of light.

```
In[52]:= Quantity[1, "SpeedOfLight"]
        UnitConvert[%]
```

Out[52]= c

Out[53]= 299 792 458 m/s

Example 13.21 Visible light covers the wavelengths 400-700 nm. Get the frequency of light corresponding to the middle of the spectrum.

```
In[41]:= UnitConvert[ c/(550. nm), Hz]
```

Out[41]= 5.45077×10^{14} Hz

Example 13.22 A certain FM radio station broadcasts at 300 MHz. Calculate the wavelength.

In[42]:= **UnitConvert$\left[\dfrac{c}{\text{300. MHz}}\right]$**

Out[42]= 0.999308 m

13.9 FOURIER SERIES

A periodic function $f(x)$ from $-L < x < L$, can be written as a sum of sine and cosine terms,

$$f(x) = a_0 + \sum_{i=1}^{\infty} a_n \cos \frac{n\pi x}{L} + \sum_{i=1}^{\infty} b_n \sin \frac{n\pi x}{L}.$$

This sum is called a Fourier series.

The integral of a sine term multiplied by a cosine term gives zero,

$$\frac{1}{L} \int_{-L}^{L} dx \, \sin \frac{n\pi x}{L} \cos \frac{m\pi x}{L} = 0,$$

while

$$\frac{1}{L} \int_{-L}^{L} dx \, \sin \frac{n\pi x}{L} \sin \frac{m\pi x}{L} = \begin{cases} 1 \text{ if } n \neq m \\ 0 \text{ if } n = m \end{cases}$$

and

$$\frac{1}{L} \int_{-L}^{L} dx \, \cos \frac{n\pi x}{L} \cos \frac{m\pi x}{L} = \begin{cases} 2 \text{ if } n = m = 0 \\ 1 \text{ if } n \neq m \\ 0 \text{ if } n = m \neq 0 \end{cases}$$

Example 13.23 Show the relationship for the above integral of the product of cosines.

```
In[44]:= $Assumptions = {L > 0, m ∈ Integers, n ∈ Integers};

        y[n_, m_] := Integrate[Cos[n π x / L] Cos[m π x / L],

        {x, -L, L}];

        y[0, 0]

Out[45]= 2 L

In[46]:= y[n, n]

Out[46]= L

In[47]:= Assuming[m ≠ n, y[n, m]]

Out[47]= 0
```

The coefficients in the Fourier series (a_n, b_n) may be found by multiplying by a sine or cosine of $(m\pi x)/L$ and then integrating. All of the terms are zero unless $m = n$,

$$a_0 = \frac{1}{2L} \int_{-L}^{L} dx \, f(x),$$

$$a_n = \frac{1}{L} \int_{-L}^{L} dx \, f(x) \cos\frac{n\pi x}{L},$$

and

$$b_n = \frac{1}{L} \int_{-L}^{L} dx \, f(x) \sin\frac{n\pi x}{L}.$$

Consider a periodic square wave (Figure 13.7).

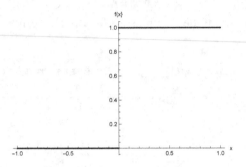

Figure 13.7 Plot of one period of a periodic square wave.

Example 13.24 Calculate the Fourier series coefficients for the square wave.

```
In[48]:= ClearAll["Global`*"];
         f[x_] = Piecewise[{{0, x < 0}, {1, x > 0}}, 0];
             1
         a₀ = ─ Integrate[f[x], {x, -1, 1}]
             2
```

$$Out[48]= \frac{1}{2}$$

```
In[49]:= Table[Integrate[f[x] Sin[n π x], {x, -1, 1}],
         {n, 1, 10}]
```

$$Out[49]= \left\{\frac{2}{\pi}, 0, \frac{2}{3\pi}, 0, \frac{2}{5\pi}, 0, \frac{2}{7\pi}, 0, \frac{2}{9\pi}, 0\right\}$$

```
In[50]:= Table[Integrate[f[x] Cos[n π x], {x, -1, 1}],
         {n, 1, 10}]
```

$$Out[50]= \{0, 0, 0, 0, 0, 0, 0, 0, 0, 0\}$$

Mathematica can quickly provide a Fourier series for any supplied periodic function.

Example 13.25 Calculate and plot the first 25, 50, 75, and 100 Fourier series terms for the square function.

In[51]:= **Table[**
  y = Re[FourierSeries[Piecewise[{{1, 0 < x < 1}}, 0],
  x, 25 i]];
  Plot[y, {x, -1, 2}], {i, 1, 4}]

Out[51]= {

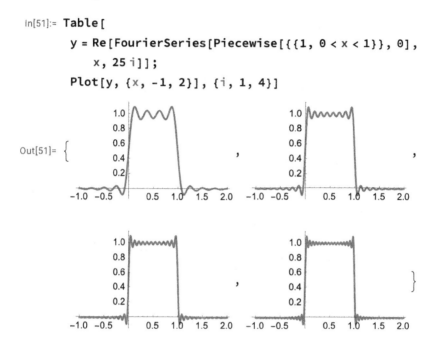

}

Figure 13.8 shows a triangular or "sawtooth" function.

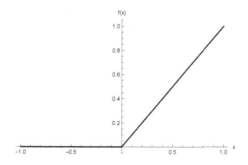

Figure 13.8 Plot of one period of a periodic sawtooth function.

Example 13.26 Calculate and plot the first 25, 50, 75, and 100 Fourier series terms for a sawtooth function.

In[52]:= `Table[`
`    y = Re[FourierSeries[Piecewise[{{x, 0 < x < 1}}, 0],`
`        x, 25 i]];`
`    Plot[y, {x, -1, 2}], {i, 1, 4}]`

Out[52]=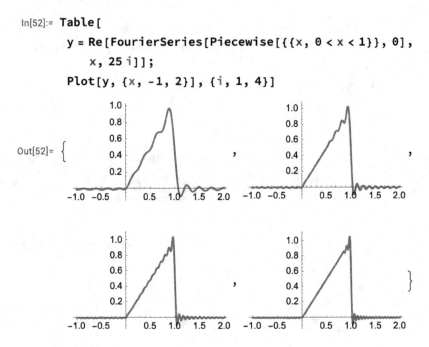

Mathematica Starter

A.1   INPUT AND TEXT CELLS

Mathematica files are called "notebooks" and have 2 types of cells: "text" cells that can only display what is written and "input" cells that are executable. The cell type is selectable from the menu bar at the top of the Mathematica notebook window. The default cell type is input. Cells may be collapsed and expanded by clicking on the bars on the right of the notebook.

Input cells are executed by typing (simultaneously)

$$\boxed{\text{SHIFT}}\ \boxed{\text{RETURN}}$$

which generates the label In[ ]:= with the output going to another input cell with the label Out[ ]= (but it is still an input cell and can be executed).

Example A.1   Calculate 1+1.

```
In[1]:=  1 + 1
Out[1]=  2
```

A semicolon after a line of code means the code will still execute but the output will be suppressed. This is a useful feature for to turn on and off when writing code to temporarily see the result of an intermediate portion of a calculation..

DOI: 10.1201/9781003481980-A

Example A.2

Set $x = 7$ and $y = 2$ and output $x + y$.

In[2]:= **x = 7; y = 2;**
   x + y

Out[3]= **9**

When a variable is set, its value may be used in other cells until cleared.

A.2 PALETTES

The menu has several extensive palettes that are useful in formatting the input. For example, there is a Writing Assistant that manages cells and fonts. This is useful for quick access to Greek letters. There is a Math Assistant that has templates for operations like division, raising to a power, summation, integration, *etc.* This makes it easy to enter an expression into an input cell in a very clean format.

Example A.3 Sum the squares of integers from 0 to 10.

In[4]:= $\displaystyle\sum_{n=0}^{10} n^2$

Out[4]= **385**

A.3 CELL DISPLAY

Cell content can be displayed in multiple forms, the most useful of which are InputForm and StandardForm. The top level "Cell" menu gives you the shortcuts to convert between displays. Note that StandardForm is selected by default, even though the result may not in StandardForm. Selecting StandardForm from the Cell menu will put the code in StandardForm. Often InputForm is easier to type, but StandardForm is nicer appearing.

Example A.4 Put the code of Ex. A.3 into InputForm.

In[5]:= **Sum[n^2, {n, 0, 10}]**

Out[5]= **385**

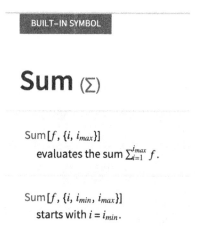

Figure A.1 This is a small part of the documentation for the function Sum.

A.4 INTRODUCTION TO FUNCTIONS

Mathematica has over 6000 functions, which always begin with a capital letter. The function Sum was used in Ex. A.4. When typing in a cell, Mathematica will give autocomplete options for existing functions. Mousing over a function gives extensive documentation for the function's use with examples. A small portion of the inline documentation on Sum is shown in Figure A.1.

The function Clear clears a variable. It produces no output.

Example A.5 Set $x = 1$ and clear x.

```
In[6]:=  x = 1
         Clear[x]
         x

Out[6]= 1

Out[8]= x
```

The function ClearAll is extremely useful to perform a global clear of everything.

Example A.6 Clear all variables.

In[9]:= `ClearAll["Global`*"]`

Some common trigonometric functions that are frequently used are Sin Cos, Tan, Cot, Sec, Csc. The following example makes use of the function List, which is executed with curly brackets { }. The List function is also used to store vectors, for example, $\{x,y,z\}$. The notation $/. \, \theta \to \pi/6$ means replace the θ with $\pi/6$ (as opposed to setting the value of θ).

Example A.7
Calculate sin, cos, tan, cot, sec, csc of $\pi/3$.

In[10]:= `{Sin[θ], Cos[θ], Tan[θ], Cot[θ], Sec[θ], Csc[θ]} /. θ →` $\dfrac{\pi}{6}$

Out[10]= $\left\{ \dfrac{1}{2}, \dfrac{\sqrt{3}}{2}, \dfrac{1}{\sqrt{3}}, \sqrt{3}, \dfrac{2}{\sqrt{3}}, 2 \right\}$

The function Simplify reduces the result algebraically.

BUILT–IN SYMBOL

List ({...})

$\{e_1, e_2, ...\}$
is a list of elements.

∨ Details

- Lists are very general objects that represent collections of expressions.

- Functions with attribute Listable are automatically "threaded" over lists

- $\{a, b, c\}$ represents a vector.

- $\{\{a, b\}, \{c, d\}\}$ represents a matrix.

- Nested lists can be used to represent tensors.

Figure A.2 This is a small part of the documentation for the function List.

Example A.8 Calculate $\sin^2 x + \cos^2 x$.

In[11]:= `Simplify[Cos[x]² + Sin[x]²]`

Out[11]= 1

It can also be written as //Simplify, placed after the code. There is a companion function, which is stronger but takes longer, called FullSimplify. Mathematica can't know how one wants the result of a calculation present, but it usually does very well. Sometimes one has to finesse the output with various substitutions. A good default is to first try the calculation without manipulation, then as a second choice add Simplify, and as a third choice add FullSimplify.

A.5   USER DEFINED FUNCTION

A user may define a function by placing an underscore after the argument, $f[x_-]$. This allows the function to be evaluated for any value of the argument.

Example A.9 Define the function $f(x) = x^2$ and evaluate it for $x = 2.5$.

In[12]:= `f[x_] = x²; f[2.5]`

Out[12]= 6.25

A.6   RESERVED NAMES

Built-in symbol names are reserved and may not be user defined., for example, D is reserved for derivative. The reserved names always begin with a capital letter. Others include, E , I, and Pi, which stand for the exponential e, imaginary i, and π.

Example A.10 Calculate $e^{i\pi} + 1$.

In[13]:= `E^I Pi + 1`

Out[13]= 0

A double equal sign makes a logical comparison.

Example A.11 Compare $E^{I\ Pi} + 1$ with $e^{i\pi} + 1$.

In[14]:= $E^{I\ Pi} + 1 == e^{i\pi} + 1$

Out[14]= True

A.7 UNITS

A unit may be obtained with the function Quantity with the common name enclosed in quotes as the argument. Mathematica is good at guessing auto-completion of what is typed. Note that everything that is not user defined begins with a capital letter.

Example A.12 Get the SI base units of meters, seconds, kilograms, kelvins, amperes, moles, and candelas.

In[20]:= {Quantity["Meters"], Quantity["Seconds"],
         Quantity["Kilogram"], Quantity["Kelvins"],
         Quantity["Amperes"], Quantity["Moles"],
         Quantity["Candelas"]}

Out[20]= $\{1\,\text{m}, 1\,\text{s}, 1\,\text{kg}, 1\,\text{K}, 1\,\text{A}, 1\,\text{mol}, 1\,\text{cd}\}$

A.8 NATURAL LANGUAGE BOX

One of the most useful features of the Wolfram language is the natural language box. This is obtained by typing $\boxed{\text{CTRL}}$ + into an input cell. One then just types into the box. If there are multiple interpretations, for example "m" could be the letter m or the abbreviation for meter, Mathematica will give a menu of interpretations to choose from. Figure A.3 shows the natural language box with "speed of light" typed into it. Clicking outside the natural

🗒 **speed of light**

Figure A.3 The natural language box is shown with "speed of light" typed into it.

Figure A.4 Result of typing "speed of light" into the natural language box.

language box evaluates it. The result is shown in Figure A.4. One simply clicks the check mark to accept the answer.

A.9 PHYSICAL CONSTANTS AND THEIR UNITS

The speed of light from the natural language box is stored as a "unit", and it appears in italics with a different shading so you can recognize the difference between a unit and a user-defined variable with the same name.

Example A.13 Get the speed of light as a physical constant and set the variable c equal to it.

In[21]:= **c = c**

Out[21]= *c*

The numerical value is displayed, together with units, using the function UnitConvert. The default units will be SI.

Example A.14 Get the numerical value of the speed of light.

In[23]:= **UnitConvert[c]**

Out[23]= 299 792 458 m/s

The units to be displayed may be specified. There are 2 ways to get a unit: (1) typing into the natural language box, and (2) using the function Quantity. Example A.15 shows both of these ways to get the speed of light in the specified units.

Example A.15 Get the speed of light in miles per second.

In[23]:= `UnitConvert`$\left[c, \dfrac{mi}{s}\right]$

`UnitConvert`$\left[$`Quantity["SpeedOfLight"]`,

$\dfrac{\text{Quantity["Miles"]}}{\text{Quantity["Seconds"]}}\right]$

Out[23]= $\dfrac{18\,737\,028\,625}{100\,584}$ mi / s

Out[24]= $\dfrac{18\,737\,028\,625}{100\,584}$ mi / s

The function N will evaluate the numerical value to the specified number of significant figures. The % symbol is a handy way to access the result of the previous calculation (%% gets the previous calculation to that, %%% gets the result of 3 calculations back, etc.). One can also use Out[n] to get the nth output.

Example A.16 Get the numerical value of the previous calculation of the speed of light to 3 significant figures.

In[24]:= `N[%, 3]`

Out[24]= 1.86×10^5 mi / s

Example A.17 Get π to 50 figures.

In[25]:= `N[`π`, 50]`

Out[25]= `3.1415926535897932384626433832795028841971693993751`

The most common derived units used in this book are the newton (N), joule (J), watt (W), and pascal (Pa).

Example A.18 Get the N, J, W, and Pa units.

In[44]:= $\{N, J, W, Pa\}$

Out[44]= $\{1\,N, 1\,J, 1\,W, 1\,Pa\}$

The default unit of UnitConvert is the SI base.

Example A.19 Convert N, J, W, and Pa into SI base units.

In[45]:= **UnitConvert[%]**

Out[45]= $\{1\,kg\,m/s^2, 1\,kg\,m^2/s^2, 1\,kg\,m^2/s^3, 1\,kg/(m\,s^2)\}$

The physical constants used in this book include the acceleration of gravity (g), universal gravitational constant (G), Boltzmann constant (k), and the mass and radius of the earth.

Example A.20 Get g, G, k, mass of earth, and radius of earth.

In[28]:= **UnitConvert**$\Big[\{$1. g, 1. G, 1. k, $\boxed{\text{Earth \small{PLANET}}}\,\big[\,mass\,\big]$,

$\boxed{\text{Earth \small{PLANET}}}\,\big[\,average\,radius\,\big]\}\Big]$

Out[28]= $\{9.80665\,m/s^2, 6.6743\times10^{-11}\,m^3/(kg\,s^2),$

$1.38065\times10^{-23}\,kg\,m^2/(s^2\,K), 5.97\times10^{24}\,kg, 6.371009\times10^6\,m\}$

Note that a one was put in front of the g, G, and k to force evaluation in decimal form.

A.10 CALCULATIONS WITH UNITS

Units are most easily assigned to a variable with the natural language box. Suppose we want to calculate how a force relates to mass times acceleration,

$$F = ma.$$

The natural language box is used to assign units (cut and paste can be a real time-saver).

Example A.21 Calculate F when $m = 26$ kg and $a = 2.8$ m/s^2 .

In[29]:= **m = 26 kg; a = 2.8 $\frac{m}{s^2}$; F = m a**

Out[29]= $72.8 \, \text{kg} \, \text{m/s}^2$

Example A.22 Get F in N.

In[30]:= **UnitConvert[%, N]**

Out[30]= 72.8 N

Mathematica is extremely useful as a calculator because it will automatically check the units of a calculation and report errors.

Example A.23 Try to get F in J.

In[31]:= **UnitConvert[F, J]**

... UnitConvert: $\dfrac{\text{Kilograms Meters}}{\text{Seconds}^2}$ and Joules are incompatible units

Out[31]= **$Failed**

A.11 SOLVING EQUATIONS

The function Solve is extremely useful for solving for a variable in an algebraic expression.

Example A.24 Solve the quadratic equation $ax^2 + bx + c = 0$ for x.

In[32]:= **ClearAll["Global`*"]; Solve[a x^2 + b x + c == 0, x]**

Out[32]= $\left\{\left\{x \rightarrow \dfrac{-b - \sqrt{b^2 - 4ac}}{2a}\right\}, \left\{x \rightarrow \dfrac{-b + \sqrt{b^2 - 4ac}}{2a}\right\}\right\}$

Note that the notation for Solve requires a double equal sign, as a single equal sign is reserved for variable assignment. The solution to Ex. A.24 is output as a list of two lists, which each contain one of the two possible solutions. It is useful to be able to pick out either of the solutions for further evaluation.

Example A.25 Choose the positive square root and evaluate it for $a = 2$, $b = 10$, and $c = 3$.

In[33]:= $(x \: /. \: \%[\![2]\!]) \: /. \: \{a \to 2, \: b \to 10, \: c \to 3\}$

Out[33]= $\dfrac{1}{4} \left(-10 + 2 \sqrt{19}\right)$

The code of Ex. A.25 says take the values of x from the previous calculation (%), pick the second one ([[2]]), and evaluate that for the specified constants a, b, and c. The arrow is obtained by typing $->$.

Solve works with any number of equations. They can be linked with &&.

Example A.26 Solve the simultaneous equations $axy = 7$ and $bx - y = 1$ for x and y.

In[34]:= $\mathbf{Solve[a \: x \: y == 7 \: \&\& \: b \: x - y == 1, \: \{x, \: y\}]}$

Out[34]= $\left\{\left\{x \to \dfrac{a - \sqrt{a}\sqrt{a + 28\,b}}{2\,a\,b}, \: y \to \dfrac{1}{2}\left(-1 - \dfrac{\sqrt{a + 28\,b}}{\sqrt{a}}\right)\right\},\right.$

$\left.\left\{x \to \dfrac{a + \sqrt{a}\sqrt{a + 28\,b}}{2\,a\,b}, \: y \to \dfrac{1}{2}\left(-1 + \dfrac{\sqrt{a + 28\,b}}{\sqrt{a}}\right)\right\}\right\}$

A.12 SERIES EXPANSION

The function Series generates a power series expansion. One can expand any function about any point, including infinity.

Example A.27 Get the first ten terms (order x^{11}) of $\sin x$ expanded about $x = 0$.

In[35]:= $\mathbf{Series[Sin[x], \: \{x, \: 0, \: 10\}]}$

Out[35]= $x - \dfrac{x^3}{6} + \dfrac{x^5}{120} - \dfrac{x^7}{5040} + \dfrac{x^9}{362\,880} + O[x]^{11}$

Example A.28 Get the first four terms of the binomial expansion.

In[47]:= `Series[(1 + x)^n, {x, 0, 3}]`

Out[47]= $1 + n x + \dfrac{1}{2} \left(-n + n^2\right) x^2 + \dfrac{1}{6} \left(2 n - 3 n^2 + n^3\right) x^3 + O[x]^4$

Example A.29 Get the first three terms of $\frac{1}{r-R}$ expanded about $r \to \infty$.

In[37]:= `Series[` $\dfrac{1}{r - R}$ `, {r, ∞, 3}]`

Out[37]= $\dfrac{1}{r} + \dfrac{R}{r^2} + \dfrac{R^2}{r^3} + O\left[\dfrac{1}{r}\right]^4$

A.13   TAKING A LIMIT

The function Limit evaluates a limit.

Example A.30 Get the limit of $\frac{\sin(x)}{x}$ as $x \to 0$.

In[38]:= $\lim\limits_{x \to 0}$ `Sin[x]` / `x`

Out[38]= 1

In Ex. A.30 the input has been put into StandardForm.

A.14   DERIVATIVES

The function D takes a derivative.

Example A.31 Set $y = x^2$ and calculate dy/dx.

In[39]:= `y = x^2; ∂_x y`

Out[39]= 2 x

In Ex. A.31 the input has been put into StandardForm.

Example A.32 Calculate the derivatives of $\sin x$, $\cos x$, and $\tan x$ with respect to x.

In[40]:= ∂_x {Sin[x], Cos[x], Tan[x]}

Out[40]= $\left\{ \text{Cos}[x], -\text{Sin}[x], \text{Sec}[x]^2 \right\}$

Example A.33 Calculate the derivative of $\sinh^{-1} x$ with respect to x.

In[41]:= ∂_x ArcSinh[x]

Out[41]= $\dfrac{1}{\sqrt{1 + x^2}}$

The function DSolve can solve a differential equation. There is one important differential equation that is encountered in beginning physics, that which describes simple harmonic motion. Simple harmonic motion occurs when there is a linear restoring force,

$$F = ma = m\frac{d^2 x}{dt^2} = -kx.$$

Example A.34 Solve the differential equation.

In[42]:= Simplify[DSolve[m ∂_t ∂_t x[t] == -k x[t] && x[0] == 0,
    x[t], t]]

Out[42]= $\left\{ \left\{ x[t] \rightarrow c_2 \, \text{Sin}\left[\dfrac{\sqrt{k}\ t}{\sqrt{m}} \right] \right\} \right\}$

A.15 INTEGRALS

The function Integrate performs an integral. The integral may be definite (limits provided) or indefinite.

Example A.35 Integrate the function x over the variable x.

In[43]:= $\int x \, dx$

Out[43]= $\dfrac{x^2}{2}$

In Ex. A.35, the input has been put into StandardForm.

Example A.36 Integrate xe^{-x} from 0 to ∞.

In[44]:= $\int_0^\infty x \, e^{-x} \, dx$

Out[44]= 1

It should be noted that indefinite integrals run faster in Mathematica, so sometimes it is a big time saver to perform the indefinite integral and substitute the limits as a second step. The integral in Ex. A.37 runs much faster when evaluated first as an indefinite integral. One can just try the definite integral first, and if that takes more than a few seconds, switch to the two-step evaluation.

Example A.37 Integrate $\dfrac{\sin\theta(r-\cos\theta)}{(\sin^2\theta+(r-\cos^2\theta)^{3/2}}$ over θ from 0 to π.

In[65]:= $Assumptions = {r > 0, r > 1};

$I[\theta\_] = \int \dfrac{\text{Sin}[\theta] \ (r - \text{Cos}[\theta])}{\left((\text{Sin}[\theta])^2 + (r - \text{Cos}[\theta])^2\right)^{3/2}} \, d\theta;$

$I[\pi] - I[0] \ // \ \text{Simplify}$

Out[65]= $\dfrac{2}{r^2}$

Integrals may be taken over multiple dimensions.

Example A.38 Calculate the volume of a ball of radius *R*.

In[45]:= **ClearAll["Global`*"];**

$Assumptions = R > 0;

Integrate[1, {x, -R, R},

$\left\{y, -\sqrt{R^2 - x^2}, \sqrt{R^2 - x^2}\right\},$

$\left\{z, -\sqrt{R^2 - x^2 - y^2}, \sqrt{R^2 - x^2 - y^2}\right\}]$

Out[45]= $\dfrac{4\pi R^3}{3}$

A.16 DRAWING OBJECTS

The function Graphics draws shapes in two dimensions. The objects themselves are specified with other functions such as Triangle, Circle, and Disk.

Example A.39 Draw a triangle and disk inside a circle.

In[47]:= **Graphics[{Gray, Triangle[], Circle[],**

Disk[{-.5, -.5}, .1]}]

Out[47]=

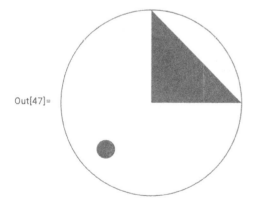

The function Graphics3D draws shapes in three dimensions. The objects themselves are specified with other functions such as Cylinder, Ball, and Cuboid.

Example A.40 Draw a cylinder, ball, and cuboid.

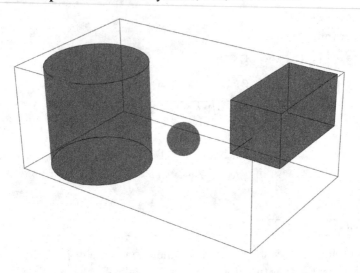

A.17 PLOTTING

Basic plots are made with the function Plot.

Example A.41 Plot $\sin x$ and $\cos(x - .2)$ from $0 \to 12$.

In[49]:= `Plot[{Sin[x], Cos[x - .2]}, {x, 0, 12},`
`PlotStyle → {Black, GrayLevel[0.7]},`
`PlotLabels → {"sin(x)", "cos(x)"}, AxesLabel → {"x"}]`

Out[49]=

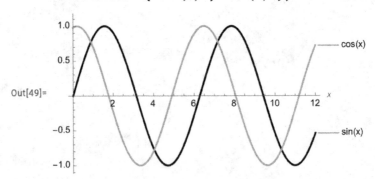

The function PolarPlot plots a curve as a function of angle.

Example A.42 Make a polar plot of sin(3θ) from $0 \to \pi$.

In[50]:= `PolarPlot[Sin[3 θ], {θ, 0, π}]`

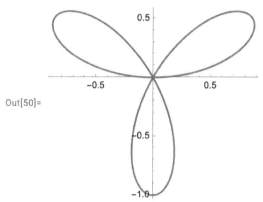

Out[50]=

the function Plot3D plots the height of a specified function $f x, y$.

Example A.43 Make a 3D plot of sin(xy) as a function of x and y.

In[51]:= `Plot3D[Sin[x y], {x, -3, 3}, {y, -3, 3}]`

Out[51]=

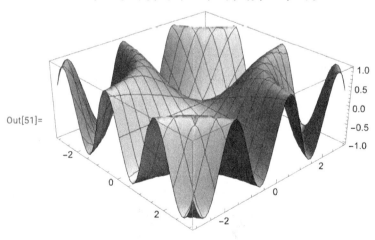

The function ParametricPlot3D makes a three-dimensional space curve.

Example A.44 Plot half of a torus.

```
In[52]:= ParametricPlot3D[
         {(2 + Cos[s]) Cos[t], (2 + Cos[s]) Sin[t], Sin[s]},
         {t, 0, π}, {s, 0, 2 π},
         PlotRange → {{-3, 3}, {-3, 3}, {-1, 1}}]
```

Out[52]=

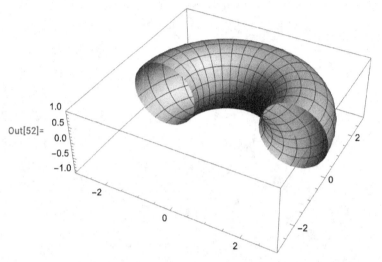

The function revolutionPlot3D generates a surface f revolution from a specified function.

Example A.45 Draw a Mexican hat.

$$
In[53]:= \text{RevolutionPlot3D}\left[\left(1 - \frac{2\,r^2}{\sigma^2}\right) \text{Exp}\left[-\frac{r^2}{2\,\sigma^2}\right] \;/.\; \sigma \to .7,\right.
$$

$$
\{r, -2, 2\}, \text{Boxed} \to \text{False}, \text{Axes} \to \text{False}\Big]
$$

Out[53]=

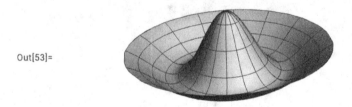

Spherical and Cylindrical Coordinates

B.1 SPHERICAL COORDINATES

Spherical coordinates are described with unit vectors $\hat{\mathbf{r}}$, $\hat{\boldsymbol{\theta}}$, $\hat{\boldsymbol{\phi}}$ that are not constants. The variable r is the distance to the origin in an arbitrary direction and has a range $0 \leq r \leq \infty$. The polar angle θ measured from the z-axis has a range $0 \leq \theta \leq \pi$. The azimuthal angle ϕ measured in the $x - y$ plane has a range $0 \leq \phi \leq 2\pi$. The unit vectors are orthogonal (see Figs. B.1 and B.2) and satisfy

$$\hat{\mathbf{r}} \times \hat{\boldsymbol{\theta}} = \hat{\boldsymbol{\phi}},$$

$$\hat{\boldsymbol{\phi}} \times \hat{\mathbf{r}} = \hat{\boldsymbol{\theta}},$$

and

$$\hat{\boldsymbol{\theta}} \times \hat{\boldsymbol{\phi}} = \hat{\mathbf{r}},$$

While spherical coordinates are convenient to describe a geometry with spherical symmetry, one must be very mindful that the unit vectors are not constant. For example, this makes derivatives non-trivial to calculate. Most of the time it is easiest to work in Cartesian coordinates (unit vectors $\hat{\mathbf{x}}$, $\hat{\mathbf{y}}$, $\hat{\mathbf{z}}$) using the spherical variables (r, θ, ϕ). These are obtained with FromSphericalCoordinates[$\{r, \theta, \phi\}$].

Example B.1 Get Cartesian coordinates (x, y, x) in terms of spherical variables r, θ, ϕ.

```
In[1]:= FromSphericalCoordinates[{r, θ, ϕ}]

Out[1]= {r Cos[ϕ] Sin[θ], r Sin[θ] Sin[ϕ], r Cos[θ]}
```

DOI: 10.1201/9781003481980-B

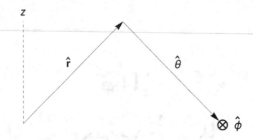

Figure B.1   Spherical-coordinate unit vectors are shown in the $r-z$ plane. The $r-z$ plane depends on the direction of $\hat{r}$.

Example B.1 says that

$$r\,\hat{r} = r\sin\theta\cos\phi\,\hat{x} + r\sin\theta\sin\phi\,\hat{y} + r\cos\theta\,\hat{z}.$$

To go in the other direction and get the spherical variables from (x, y, x), for example, use ToSphericalCoordinates[$\{x, y, x\}$].

Example B.2 Get spherical coordinates (r, θ, ϕ) in terms of Cartesian variables x, y, z.

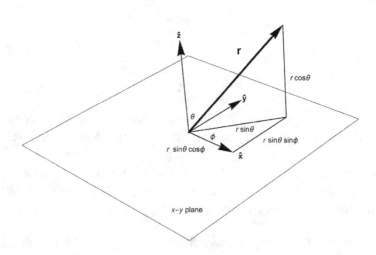

Figure B.2   In spherical coordinates (r, θ, ϕ), the polar angle (θ) is the angle between the vector **r** and the z-axis and the azimuthal angle (ϕ) is the angle in the $x-y$ plane.

In[2]:= `ToSphericalCoordinates[{x, y, z}]`

Out[2]= $\left\{ \sqrt{x^2 + y^2 + z^2}, \; \text{ArcTan}\left[z, \sqrt{x^2 + y^2}\right], \; \text{ArcTan}[x, y] \right\}$

Example B.2 says that

$$x\,\hat{\mathbf{x}} + y\,\hat{\mathbf{y}} + z\,\hat{\mathbf{z}} = \sqrt{x^2 + y^2 + z^2}\,\hat{\mathbf{r}} + \tan^{-1}\left(\frac{\sqrt{x^2+y^2}}{z}\right)\hat{\boldsymbol{\theta}} + \tan^{-1}\left(\frac{y}{x}\right)\hat{\boldsymbol{\phi}}$$

To get a Cartesian unit vectors in terms of spherical unit vectors, use TransformedField["Cartesian"→ "Spherical", f, $\{x, y, x\} \to \{r, \theta, \phi\}$], where f is the vector field to be transformed. Since the calculation is repetitive, several shortcuts can be used. The symbol # is an abbreviation for the function Slot which makes a substitution for the function after the &, and /@ is an abbreviation for the function Map. The function IdentityMatrix[n] is a matrix of unit vectors for n dimensions. The function Column outputs the vectors in a column.

Example B.3 Get Cartesian unit vectors $\hat{\mathbf{x}}, \hat{\mathbf{y}}, \hat{\mathbf{z}}$ in terms of spherical unit vectors $\hat{\mathbf{r}}, \hat{\boldsymbol{\theta}}, \hat{\boldsymbol{\phi}}$ and angles.

In[3]:= `TransformedField["Cartesian" → "Spherical", #,`
`            {x, y, z} → {r, θ, φ}] & /@ IdentityMatrix[3] // Column`

Out[3]= `{Cos[φ] Sin[θ], Cos[θ] Cos[φ], -Sin[φ]}`
`{Sin[θ] Sin[φ], Cos[θ] Sin[φ], Cos[φ]}`
`{Cos[θ], -Sin[θ], 0}`

Example B.3 says that

$$\hat{\mathbf{x}} = \sin\theta\cos\phi\,\hat{\mathbf{r}} + \cos\theta\cos\phi\,\hat{\boldsymbol{\theta}} - \sin\phi\,\hat{\boldsymbol{\phi}},$$

$$\hat{\mathbf{y}} = \sin\theta\sin\phi\,\hat{\mathbf{r}} + \cos\theta\sin\phi\,\hat{\boldsymbol{\theta}} + \cos\phi\,\hat{\boldsymbol{\phi}},$$

and

$$\hat{\mathbf{z}} = \cos\theta\,\hat{\mathbf{r}} - \sin\theta\,\hat{\boldsymbol{\theta}}.$$

Example B.4 Get spherical unit vectors $\hat{\mathbf{r}}, \hat{\boldsymbol{\theta}}, \hat{\boldsymbol{\phi}}$ in terms of Cartesian unit vectors $\hat{\mathbf{x}}, \hat{\mathbf{y}}, \hat{\mathbf{z}}$ and angles.

In[4]:= **Simplify[**
    TransformedField["Cartesian" → "Spherical", #,
        {x, y, z} → {r, θ, φ}] & /@
    TransformedField["Spherical" → "Cartesian", #,
        {r, θ, φ} → {x, y, z}], {r > 0, 0 < θ < π}] & /@
    IdentityMatrix[3] // Column

        {Cos[φ] Sin[θ], Sin[θ] Sin[φ], Cos[θ]}
Out[4]= {Cos[θ] Cos[φ], Cos[θ] Sin[φ], -Sin[θ]}
        {-Sin[φ], Cos[φ], 0}

Example B.4 says that

$$\hat{\mathbf{r}} = \sin\theta\cos\phi\,\hat{\mathbf{x}} + \sin\theta\sin\phi\,\hat{\mathbf{y}} + \cos\theta\,\hat{\mathbf{z}},$$

$$\hat{\boldsymbol{\theta}} = \cos\theta\cos\phi\,\hat{\mathbf{x}} + \cos\theta\sin\phi\,\hat{\mathbf{y}} - \sin\theta\,\hat{\mathbf{z}},$$

$$\hat{\boldsymbol{\phi}} = -\sin\phi\,\hat{\mathbf{x}} + \cos\phi\,\hat{\mathbf{y}}.$$

B.2   CYLINDRICAL COORDINATES

Cylindrical coordinates keep the z-axis fixed, using unit vectors $\hat{\mathbf{r}}$, $\hat{\boldsymbol{\phi}}$, $\hat{\mathbf{z}}$. The variable r is the distance to the z-axis and has a range $0 \le r \le \infty$. The azimuthal angle ϕ measured in the $x-y$ plane has a range $0 \le \phi \le 2\pi$. The unit vectors are orthogonal (see Figs. B.3 and B.4) satisfy

$$\hat{\mathbf{r}} \times \hat{\boldsymbol{\phi}} = \hat{\mathbf{z}},$$

$$\hat{\boldsymbol{\phi}} \times \hat{\mathbf{z}} = \hat{\mathbf{r}},$$

and

$$\hat{\mathbf{z}} \times \hat{\mathbf{r}} = \hat{\boldsymbol{\phi}},$$

Example B.5 Get Cartesian coordinates (x, y, x) in terms of cylindrical variables r, ϕ, z.

In[5]:= **CoordinateTransform["Cylindrical" → "Cartesian", {r, φ, z}]**

Out[5]= {r Cos[φ], r Sin[φ], z}

Example B.5 says that

$$r\,\hat{\mathbf{r}} + z\,\hat{\mathbf{z}} = r\sin\phi\,\hat{\mathbf{x}} + r\sin\phi\,\hat{\mathbf{y}} + z\,\hat{\mathbf{z}}.$$

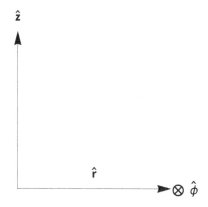

Figure B.3 Cylindrical-coordinate unit vectors are shown in the $r - z$ plane. The $r - z$ plane depends on the direction of $\hat{r}$.

Example B.6 Get cylindrical coordinates (r, ϕ, z) in terms of Cartesian variables x, y, z.

In[6]:= **CoordinateTransform["Cartesian" → "Cylindrical", {x, y, z}]**

Out[6]= $\left\{ \sqrt{x^2 + y^2}, \text{ArcTan}[x, y], z \right\}$

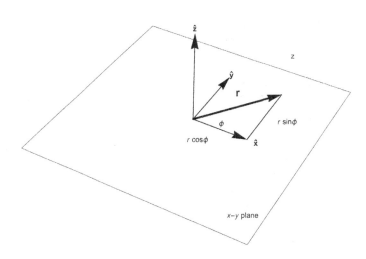

Figure B.4 In cylindrical coordinates (r, ϕ, z), the azimuthal angle (ϕ) is the angle in the $x - y$ plane.

Example B.6 says that

$$x\,\hat{\mathbf{x}} + y\,\hat{\mathbf{y}} + z\,\hat{\mathbf{z}} = \sqrt{x^2 + y^2}\,\hat{\mathbf{r}} + \tan^{-1}\left(\frac{y}{x}\right)\hat{\boldsymbol{\phi}} + z\,\hat{\mathbf{z}}$$

Example B.7 Get Cartesian unit vectors $\hat{\mathbf{x}},\hat{\mathbf{y}},\hat{\mathbf{z}}$ in terms of cylindrical $\hat{\mathbf{r}},\hat{\boldsymbol{\phi}},\mathbf{z}$.

```
In[7]:= TransformedField["Cartesian" → "Cylindrical", #,
           {x, y, z} → {r, φ, z′}] & /@ IdentityMatrix[3] // Column

         {Cos[φ], -Sin[φ], 0}
Out[7]= {Sin[φ], Cos[φ], 0}
         {0, 0, 1}
```

Note that in Ex. B.7 the coordinate z could not be used twice so z' was used for cylindrical. Example B.7 says that

$$\hat{\mathbf{x}} = \cos\phi\,\hat{\mathbf{r}} - \sin\phi\,\hat{\boldsymbol{\phi}},$$

$$\hat{\mathbf{y}} = \sin\phi\,\hat{\mathbf{r}} + \cos\phi\,\hat{\boldsymbol{\phi}},$$

and

$$\hat{\mathbf{z}} = \hat{\mathbf{z}}'.$$

Example B.8 Get cylindrical unit vectors $\hat{\mathbf{r}},\hat{\boldsymbol{\phi}},\mathbf{z}$ in terms of Cartesian $\hat{\mathbf{x}},\hat{\mathbf{y}},\hat{\mathbf{z}}$.

```
In[8]:= FullSimplify[
             TransformedField["Cartesian" → "Cylindrical", #,
                {x, y, z} → {r, φ, z′}] & /@
             TransformedField["Cylindrical" → "Cartesian", #,
                {r, φ, z′} → {x, y, z}], {r > 0, 0 < θ < π}] & /@
         IdentityMatrix[3] // Column

         {Cos[φ], Sin[φ], 0}
Out[8]= {-Sin[φ], Cos[φ], 0}
         {0, 0, 1}
```

Example B.8 says that

$$\hat{\mathbf{r}} = \cos\phi\,\hat{\mathbf{x}} + \sin\phi\,\hat{\mathbf{y}},$$

$$\hat{\boldsymbol{\phi}} = -\sin\phi\,\hat{\mathbf{x}} + \cos\phi\,\hat{\mathbf{y}},$$

$$\hat{\mathbf{z}}' = \hat{\mathbf{z}}.$$

Index

Printed in the United States
by Baker & Taylor Publisher Services